THE GOOD LIVING GUIDE TO

BEEKEEPING

THE GOOD LIVING GUIDE TO

BEEKEEPING

SECRETS OF THE HIVE, STORIES FROM THE FIELD, AND A PRACTICAL GUIDE THAT EXPLAINS IT ALL

DEDE CUMMINGS

FOREWORD BY SCOTT WILSON

Good Books

New York, New York

Good Books books may be purchased in bulk at special discounts for sales promotion, corporate gifts, fund-raising, or educational purposes. Special editions can also be created to specifications. For details, contact the Special Sales Department, Good Books, 307 West 36th Street, 11th Floor, New York, NY 10018 or info@skyhorsepublishing.com.

Good Books is an imprint of Skyhorse Publishing, Inc.®, a Delaware corporation.

Visit our website at www.goodbooks.com.

10 9 8 7 6 5 4 3

Library of Congress Cataloging-in-Publication Data is available on file.

Cover design by Jenny Zemanek

Print ISBN: 978-1-68099-059-1
Ebook ISBN: 978-1-68099-110-9

Printed in the United States of America

Permission gratefully received to use recipes from BeeMaid, Eating Well, Food Network, and the National Honey Board.

Contents

Photo of the author by Catherine Dianich Gruver.

Foreword

The plight of the honeybee and other pollinators has been the focus of much media attention in recent years. Issues such as Colony Collapse Disorder (CCD)—where entire colonies of honeybees would for no known reason vacate a hive—have brought the threat to the continued existence of the honeybee to the world's attention. This focus has led to the creation of the Saving America's Pollinators Act (H.R.1284), which directs the Environmental Protection Agency to take certain actions related to pesticide that may affect pollinators. Beekeepers are on the forefront of this action.

Beekeepers share a relationship and a responsibility with and to the honeybee, one that is important from the most experienced beekeeper to the apprentice. Beekeepers also have a fiducial responsibility to learn and practice the art of managing healthy honeybees. As such, they tend to seek out new thoughts, ideas, and practices from other beekeepers in the hopes to improve their skills and refine the art of beekeeping.

You can usually find a group of beekeepers huddled in conference rooms buzzing about the latest research, eagerly lapping up every piece of sweet information. They peruse websites and newsgroups hoping to glean an insight into a new piece of information. They tend to be social people with a common goal . . . to manage healthy, productive honeybees. The honey bee is also a social being. Each bee in the hive, depending on gender and age, has specific hive duties. The males (drones) are responsible for providing genetic diversity and their sole job is to mate with a queen; the queen's sole responsibility is to lay eggs; and the worker bee (female), throughout her lifetime, will play multiple roles within the hive ranging from feeding the young and guarding the hive, to foraging for pollen or nectar. The caste of this system is clear and work roles are well defined. They labor as a society to achieve the growth of the colony.

There could never be enough books written about honeybees. Just as each hive has its own personality so does each book written about bees. Dede Cummings provides a book with an approach that is unique in its application. Interspersed throughout the book are amusing stories about beekeepers as well as their individual techniques with beekeeping. Not only will the reader learn about necessary concepts like smokers, hive tools, or veils, but will also have the benefit of reading about actual examples of local beekeepers relating their successes and failures.

Cummings has bridged the gap between a typical textbook and "how to" book by providing the reader with pertinent information coupled with humorous, but applicable, real-world examples. The symbiotic relationship between beekeeper and honeybee is displayed by the words of the beekeepers as they discuss why they love keeping bees. The novice beekeeper will gain a foundational understanding of the basic requirements for beginning this fascinating hobby. The seasoned beekeepers will smile warmly as they recall similar experiences from their own work.

Thank you to Dede for preparing this book and giving us another reason to keep and love bees.

—Scott Wilson, Vermont Beekeeper at Heavenly Honey Apiary

INTRODUCTION

"Weeds are flowers, too, once you get to know them."
—A. A. Milne, from *Winnie the Pooh*

As a girl growing up in New England, I remember being fascinated by bees and honey and *Winnie the Pooh*. Down by the well on the property we rented in the summertime, I would curl up with the book and, uninterrupted by the voices of my younger sisters, was free to read and linger on each page. Sometimes I put the book down and explored by the brook.

It was there, in southern Rhode Island, when I first came upon a bee. It was late spring, and the solitary creature was slowly making its way from flower to flower. There were willow trees bending to the water. I pulled on one of the long tassels and watched as the bee busied itself with the blooms. There was an old orchard of apple trees just beyond the brook. I was fascinated. I sat and watched this industrious creature, curious as to why it was alone when I had read a book at school about hives and colonies.

I have since learned that the single bee I saw that day was a "native" bee. The solitary bees were here before the Native Americans lived and long before the settlers arrived, bringing with them the honeybee. The solitary bee was in fact native to North America; the honeybee was native to Europe (also Africa and Asia).

This early fascination with bees has led me to my new passion: learning about the kind of bees that make honey. I use honey every day. I coax a spoonful into my English breakfast tea every morning, spoon it onto a banana for a snack, use it for baking. I gave up refined sugar a few years ago and now only use honey when I want a little sweetener.

I did this for health reasons. Honey is a monosaccharide, a much healthier and more easily digestible alternative to trisaccharides like table sugar. And it is so pretty—it is golden in my glass Ball jar that I use for refilling. I keep it on a counter in my kitchen and sometimes the sun hits it just right and it almost glows.

PART I

ALL ABOUT BEES

"The keeping of bees is like the direction of sunbeams."
—Henry David Thoreau

INTRODUCING . . .
THE AMAZING HONEYBEE

"[The bee is] the venerable ancestor to whom we probably owe most of our flowers and fruits (for it is actually estimated that more than a hundred thousand varieties of plants would disappear if the bees did not visit them) and possibly even our civilization . . ."

—Maurice Maeterlinck

Bee Culture: Queens, Workers, Drones, and Their Roles

Before we begin, it's important for you to understand the culture of bees—or at least the best that you can. It will help you to become a better beekeeper and will make the process, with all its ups and downs, all the more enjoyable.

Honeybees live in a female-dominated society, and are surprisingly social. A colony will generally include one queen bee (the fertile egg-laying

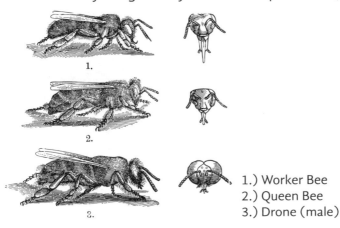

1.) Worker Bee
2.) Queen Bee
3.) Drone (male)

female), a small population of drone bees (the lone and short-lived males), and a large number of sterile female worker bees. Siblings share three-quarters of their genome, which helps to explain their cooperative style of living.

Because she is the only sexually developed female in the colony, the queen's primary function is reproduction. And she will do her job, all day, every day, for all of the warmer months of the year. On an energetic day, she can lay around two thousand eggs. This is only possible with the support of her attendants, who clean and preen and primp her, all the while feeding her the bee equivalent of energy bars: royal jelly. And thanks to this special delicacy, she is able to produce more than a million eggs in her lifetime.

A queen's second role, beyond incredible reproductive success, is to produce queenly pheromones. These chemical signals serve as a "social glue" to unify the colony, and affect mating behavior, ensure maintenance of the hive, and inhibit ovary development in the female worker bees.

THE DIFFERENCES BETWEEN BUMBLEBEES AND HONEYBEES

There is often confusion about the differences between bumblebees and honeybees. This chart explains some of the differences. (Used with permission from the Bumblebee Conservation Trust.)

Bumblebee	Honeybee
Fat and furry appearance.	Smaller and slim appearance, like a wasp.
24 different species of bumblebee in the UK.	Only one species of honeybee in Europe.
Different species have different lengths of tongue. This means they feed from different shaped flowers.	All honeybees have short tongues so they prefer open flowers.
Bumblebees live in nests with 50–400 bees.	Honeybees live in hives of up to 50,000–60,000 bees.
Only the queen hibernates, in a hole in the ground.	The queen and many of her daughters live in the hive all year
The queen lives for one year, but the other bumblebees only live for a few months.	The queen can live for three–four years.
They live in the wild, e.g., in gardens and the countryside.	Most honeybees are looked after by beekeepers, but there are some wild colonies.
Bumblebees only make small amounts of a honey-like substance to eat themselves.	Honeybees make lots of honey, which beekeepers can harvest to eat or sell.
Bumblebee populations are declining due to a shortage of flowers to feed from and places to nest in the countryside.	Honeybees are mainly declining due to diseases and mites, such as the Varroa mite.
They can sting more than once but only sting if aggravated.	Honeybees die after they have stung as their stinger is barbed and sticks in the skin.
Don't dance but may communicate by passing pollen between worker bees.	Use a "waggle dance" to communicate, passing on information about flower locations.

Wild bumblebees face many threats including habitat alteration, pesticide use, management practices, and pathogens that their honey-making counterparts, the honeybee, face. According to The Xerces Society for Invertebrate Conservation, in the late 1990s, bee biologists started to notice a decline in the abundance and distribution of several wild bumblebee species. Five of these species (western bumblebee, rusty patched bumblebee, yellow banded bumblebee, and the American bumblebee) were once very common and important crop pollinators over their ranges.

Bees in Ancient Times

Images of humans collecting honey from bees date back thousands of years. The ancient Egyptians actually attempted to domesticate bees approximately 4,500 years ago. As far back as the Paleolithic age, gathering honey has been noted in art and culture. Traditions of "honey hunts"—gathering honey from bees in the wild—and reverence for bees is apparent in many cultures.

The Bee Crisis and Why It Matters

Bees can be sold or rented in order to be used for crop pollination, like almonds in California which are almost entirely dependent on the bees for pollination, but colony collapse and other problems, such as tired bees, have made world headlines.

Commercial apiarist David Hackenburg was the first beekeeper to notice what was to be called Colony Collapse Disorder (CCD). Along with Dave Mendes, a beekeeper from Florida, these two alerted the rest of the world's beekeepers and the media about eight years ago to the modern-day plight of the bees. The world listened, but the commercial growers were at a loss to figure out why their once lucrative business and practices were failing.

Scientists are not sure, but something is compromising the bees' immune system. Scientists at Penn State, under the director Dennis van Engelsdorp, State Apiarist for Pennsylvania's Department of Agriculture, are anxiously trying to determine the cause of colony collapse disorder. Author and food activist Michael Pollan says that the California Central Valley is one of the growing centers of the world for almonds, and bees are being shipped from as far away as Australia to help pollinate the trees. This type of monoculture is not only stressful on the bees, it is dependent on pesticides and artificial practices to maintain the colonies.

"There is one masterpiece, the hexagonal cell, that touches perfection. No living creature, not even man, has achieved, in the centre of his sphere, what the bee has achieved in her own: and were some one from another world to descend and ask of the earth the most perfect creation of the logic of life, we should needs have to offer the humble comb of honey."
—Maurice Maeterlinck, *The Life of the Bee*, 1924

THE ANATOMY OF HONEYCOMB

The near-perfect shape of hexagonal honeycomb has perplexed and awed scientists and beekeepers for centuries. Just recently, in 2013, a study done by engineers in the UK and China has revealed that the holes in the comb start off as circular and then take a hexagonal shape when the wax hardens. Below, I outline the biological process of building honeycomb.

1. Young adult bees secrete a substance known as beeswax from glands on their abdomens.
2. Adult bees harvest this off the young and process it in their mouths.
3. After processing, the adult bees begin to shape the material into a comb.
4. The comb will eventually store bee larvae, honey, nectar, and pollen.

The Future of Beekeeping

Holistic and organic beekeepers take a different approach. According to Günther Hauk, author of *Toward Saving the Honeybee*, a firm grasp of the most beneficial practices in raising and maintaining hives is necessary. Mechanized, artificial queen breeding and artificial insemination are practices that are common with commercial beekeepers. According to Hauk, and other bee leaders like Dee Lusby, founder of the organic beekeeping group, the migratory beekeepers need to focus on the natural state in which the bees thrive.

Genetically modified farming and treating crops with pesticides are leading us down a road of no return. The system of agriculture in the United States and many parts of the world is clearly not working due to unsustainable practices like monoculture farming. Rather than continue to expose bees to pesticides, it is time for a major overhaul of current practices. Corn and soybeans that are treated with systemic pesticides are examples of what Michael Pollan calls "cheap food."

This book is one small offering and plea to avoid genetically modified growing of crops and focus on the natural way to grow food. Perhaps it won't be too late to save the bees. In a movement that is spreading around the world, it is entirely appropriate to begin keeping bees in one's own backyard. As Henry David Thoreau said, "There are certain pursuits which, if not wholly poetic and true, do at least suggest a nobler and finer relation to nature than we know. The keeping of bees, for instance."

Getting Started with Backyard Beekeeping

Photo of the author by Catherine Dianich Gruver.

"The bee is more honored than other animals, not because she labors, but because she labors for others."
— Saint John Chrysostom (Archbishop of Constantinople, 347–407)

There is no doubt that bees are important. They are responsible for every third bite of food we humans take. Many people call beekeeping a hobby, but really, it is a lifestyle that is both demanding and beautiful, allowing us to earn earth's sweetest ambrosia and best wax in the process. There is no way one can just jump into the trade. Talk to beekeepers in your area, search the vast internet, read a good old-fashioned book, do whatever it takes to educate yourself before you take the final plunge of deciding that beekeeping is for you.

Make plans before you order or catch your bees, figure out your game plan of where they will go and what hive you think will suit your needs. Learn all of their needs and the adjustments you will need to make if ever something were to go wrong. They are self-sustaining, yes, but cannot abundantly thrive without a little tender love and care. Understand all there can be understood from bees. But most of all, try not to think of your backyard bees as a burden. Yes, there will be problems such as losing most of your bees to varroa mites as happened to Bill Mares, an experienced beekeeper of forty years, or spending the night in a field due to faulty truck mechanics because you had to move bees at dusk (nighttime being the best time to move bees) like Jennings Berger, a skillful beekeeper of thirteen years. Heck, you may even become the stinging target of a swarm returned to their hive, which happened on a hot summer day to Mickey Rowland, a seasoned beekeeper of twelve years who lives among his bees in Cape Cod. The key is not to give up. Backyard beekeeping is not—nor will it ever be—easy. However, it will be rewarding and always seems to be a learning experience unlike any other.

JENNINGS BERGER

Jennings Berger grew up on a ranch in Oklahoma, where his mom was an author and his dad was a professional rodeo cowboy, cattleman, and horse trainer. His parents divorced when he was eleven, his dad went to south Texas, and his mother moved to Brooklyn, NY. His sisters and he went to school in New York and spent summers in Texas. After graduating from high school, he attended Castleton State College, completing his degree in Natural Science. In 2001 he became interested in beekeeping. For the past fourteen years, he has evolved from a hobby beekeeper to owning his own small business, Stone Arch Apiaries. He produces honey and beeswax, as well as raises queens. Jennings is a member of the Vermont Beekeeping Association. In 2010, he was selected by Mike Palmer and Bill Mares to

participate in a grant funded queen-rearing program. He is now part of a small group of well-trained queen breeders that ensure a supply of good local queens. In addition to his beekeeping business, he is currently the lead mechanical installer at a solar company. He lives in Brookline, Vermont with his wife, son, and baby on the way.

How long have you been keeping bees?
I have been keeping bees for thirteen years.

What inspired you to begin as a beekeeper?

My wife had an interest in soap and candle making and ordered a bunch of catalogues. The beekeeping suppliers are the primary suppliers of this stuff. I spent a winter perusing their catalogues,  and decided beekeeping would be fun to try. I quickly became addicted. Honeybees are amazing creatures that we humans will probably never fully understand. There are a select few true super organisms on earth, and honeybees are one of them. To answer your question, I started keeping bees on a whim, just thought it would be neat to try it out. Beekeeping quickly became an amazing adventure of discovery and an endless learning experience that will last the rest of my life. I have no illusions of ever knowing everything about bees. I equate it to people who love golf—there is room for a lifetime of refinements and improvement.

Any tips for beginners?

Beginners should try to find a mentor; someone who has bees and is willing to give you some pointers helps a lot. Join a local beekeeping club. I did not have a mentor; I had to figure most things out for myself, but when I started going to meetings the whole beekeeping world opened up. Go to meetings, enjoy the lectures, visit with your fellow beekeepers—you will learn a lot, and gain valuable resources. Most of all, relax, enjoy the bees. Manipulating a hive is an art, knowing what you are seeing is a science.

If your goal is to run a beekeeping business, I would say diversify your products. In general, I would say pay attention to current research. Keep up on the politics surrounding bees. Keep a subscription to *American Bee Journal* and/or *Bee Culture* magazine. Go to your local beekeeping meetings.

Please relate a memorable experience you had as a beekeeper—humorous, instructional, tragic, or all of the above.

Bees should be moved just after dark, or before daylight when all the field bees are in the colony. One spring night, I was delivering hives for pollination at a second home that is very empty and very remote, and the battery in my truck died. It was about 9 p.m., I had no cell service, and it was miles to the nearest inhabited residence. So I waited for a while, and tried the truck again. Nothing! I hung around for a while messing with it, thinking that if I give it ten minutes to rest it would start. All the while I was really just stalling the inevitable walk into the darkness.

Eventually, there was no choice but to strike out. So I walked up through this huge field, around the house and out to the road. Hmm, which way do I go? Well I knew that there hadn't been signal for miles in the direction I came from, because I was using the GPS in my phone to find the place. So I ventured farther into the unknown. I started walking on a tiny dirt road I had never been over.

After a couple of miles, I passed two empty houses and endured countless rustlings in the bushes. I finally got enough signal to make calls. I called home to let my wife know what was happening, and then called AAA. AAA said someone will be there in one hour. I explained my situation to them in depth, and they assured me they would be there. They told me to return to my vehicle. So I walked all the way back, but I had to sit by the road. My truck was out in the middle of the fields—they would never find me if I didn't wait by the road. I picked out a good stump to sit on across from the driveway and *waited* in the dark, *forever*. Two and a half hours later, I walked back and called them again. The nice lady at dispatch said, "Your call just came in, we will have someone there in one hour. Yes, return to your vehicle." After I freaked out on her for a few minutes, I took the long dark walk for the fourth time. I returned to my stump and waited. I heard the truck long before the headlights came into view. A friendly fellow with a jumpstart pack. We walked down through the fields talking. The truck started right up. After giving him a ride up to his truck and filling out the necessary paperwork, I was on my way. I got home around 6 a.m., just in time to call in sick to work and climb into bed.

What products do you make?
I currently sell honey, wax, and queen bees.

How many hives do you recommend someone begin with?
Two to four. It's good to have more than one so you have a comparison.

What beehive do you use/ like?
I use 10-frame Langstroth equipment. Brood chambers are 2 deep, with medium (6⅝) supers.

Where do you get your bees?
There are many places to buy nucleus colonies and package bees. If you already have bees you can buy queens and split your colonies. If you have plenty of bees and are ambitious, you can raise your own queens and make your own new colonies.

Where is a good spot to put a beehive?
In northern areas bees should be located in sunny locations—a sunny, dry location with water nearby, and good wind protection. A hedgerow or other windbreak is good. Stay away from frost pockets and low foggy areas. If there are going to be large numbers of hives you should be able to drive to them. I can back a truck right up to all of my bee yards. I paint all my hives darker colors, not white. In northern climates the long cold winter is more of a concern than the summer heat. Southern yards are placed in areas with afternoon shade, and painted white. The most important thing is forage. If there are no honey plants the bees will not make any honey.

Choosing a Spot

When it comes time to figure out where to place your hive, there are a few things to consider first—one being water accessibility. Bees can fly a maximum of three miles, but if they have to fly that far to get water, they will grow tired, and honey production will decrease. The water source has to be constant and close to their hive. Naturally, bees are attracted to running water like a stream or brook, but still water such as an open basin or pond is better and safer due to the lack of current. It is important to note that bees cannot hover over the water while trying to obtain a drink, and they will drown unless they have something to land on. To help them out, place rocks in the water that break the surface of the filled basin or make sure there are spots on a pond's edge for the bees to stand on.

You also have to make sure there are sustainable quantities of pollen and nectar available. The closer the hive can be to the garden the better because the busy bees will take up less time traveling and more time creating their sticky sweet honey. If you don't have a large selection of flowers in your garden, species such as clovers and dandelions are really helpful to have somewhere nearby on your property.

Sunlight is important when it comes to honey production, so make sure your hive is in direct sunlight from sunup to sundown. Because bees are active all day, if they were to be in the shade when the sun rose, they wouldn't start their day until the golden light reached their hive. Facing the hive toward the sunrise is a good wake-up call for the workers.

Although you may think your hive is sturdy, wind force is easily (and almost always) underestimated. To ensure the safety of the hive, create a windbreak. There are probably already natural or prestanding windbreaks on your property, such as tree lines, fence posts, tall bushes, or the side of a building. If none of these fit your fancy or are not available to you, you can always buy a windbreak.

Now, just because bees are attracted to running water does not mean they enjoy a damp hive. The location you choose must have proper drainage which usually means you should place your hive on a small slope.

Last, but definitely not least, you should use the well of common sense you harbor when picking a spot. If there are other hives in the area and not enough pollen and nectar sources, competition will define honey supply and quality. Also, bees don't really care who is in their way, so try and keep hives away from travel paths and swing sets.

HEAVENLY HONEY APIARY,
owned and operated by Scott and Valarie Wilson

For more about Scott and Valarie Wilson, please see Success Stories on page 123.

How long have you been keeping bees?
We have just begun our ninth season of beekeeping.

What inspired you to begin as a beekeeper?
Fear and fun. Our only child had begun ninth grade. Valarie was sensing that her days were numbered as a family unit of three. That's the fear part—being empty nesters. Now the fun part was that we had just purchased a house that butts up to an apple orchard. So the day that Valarie heard a radio announcement for a beginner's beekeeping class her ears perked up. Scott, having heard the same advertisement, was amazed when he asked Valerie this question: "What do you want our next hobby to be?" Well, as you probably guessed, she said "beekeeping." They took that beekeeping class together, and the rest is history.

Any tips for beginners?
Always have an EpiPen with you (it doesn't matter that they cost a lot of money!), and always leave more honey for your bees going into winter than what the books say, and join your local beekeeping association—they are an invaluable tool, maybe more so than your hive tool!

Tips for experienced beekeepers?
Always have two EpiPens with you. Don't stop learning. What worked last year may not work this year. Keep a colony journal to refer back to (because your memory may get confused when you have more than one hive).

Please relate a memorable experience you had as a beekeeper—humorous, instructional, tragic, or all of the above.

Humorous:
A few years back, Scott went to check if a caged queen installed into a Nuc made from a weaker hive had released. "Hmmm," he said as he lifted the inner cover. 'Where is the queen cage?' Turns out, he installed the queen into the wrong Nuc—but into one that, of course, already had a queen. The newly released queen was nowhere to be found. This is what happens when one is rushed, hot, and fatigued.

Tragic:
This is rare, which bears repeating—this is rare. During our fourth season of beekeeping and after many stings, Valarie received a bee sting that had us rushing to the hospital to be treated for anaphylactic shock. Now you know why we insist on having an EpiPen on hand at all times. We were seasoned beekeepers, both working the hives and loving every minute of it. However, this non-systemic reaction changed things up. We now call Valarie the 'inside beekeeper' and Scott handles all the actual beekeeping of the colonies. (*Thankfully the only tragedy in this story is that Scott and Valarie don't get to work the hives together, but they still get to run their side-line beekeeping business as a team.*)

What products do you make?
Raw honey (not filtered and not pasteurized) in its liquid and
   crystallized states
Beeswax
Beeswax candles
Raw bee pollen
Raw bee propolis
We also offer honey extraction services

How many hives do you recommend someone begin with?
Three. This gives you a good comparison on what's normal, etc.
Going into fall, if you have two weak hives you can merge them and
make a robust colony, etc.

What beehive do you use/like?
Langstroth. Although, we are trying out other hives such as top bar
to get a feel for other structures.

Where do you get your bees?
We make our own walk-away splits and develop our own nucleus
colonies (or "Nucs" for short) for overwinter. When we have to
purchase, our aim is to buy our Nucs (preferably local with a Vermont
queen) and packaged bees from Betterbee.

Where is a good spot to put a beehive?
We encourage beekeepers to select a spot that faces a southeasterly
direction. This is to take advantage of the early morning sun. A spot
that has a tree line to the north is helpful to cut down on the cold
winter winds. We avoid high spots such as the tops of hills or low
spots where cold air may settle.

How do you attract bees?
This is Vermont; we don't have any trouble attracting honeybees with
all of our native wildflowers, annuals, and perennials. However, every
year we invest in more plants for our gardens, such as lupine, sedum,
and mint.

What do you do when bees or the whole hive become sick?
There are a variety of bee health issues that can affect a hive. The most devastating is called American Foul Brood. It is the one brood disease that typically requires the incineration of the bees and the hive. It is so destructive that it can wipe out an entire apiary and, due to its virulent nature and the honeybees' ability to fly many miles, the potential for other beekeepers, hives to be contaminated is high. One of the most pressing pest issues is a mite called Varroa Destructor. It is a parasite that gestates inside the cells or drone comb, damaging the bee there. Then, once it emerges, it attaches to a honeybee, piercing the bee's exoskeleton and living off the bee's hemolyph (bee blood). The hole then becomes a portal for other hive diseases to enter the bee. We try to manage the mite loads by using naturally occurring chemicals such as thyme oil extract and formic acid. These help to control the mite population in the hopes that the bees can stay strong and disease-free.

What flowers do you find bees are most attracted to?
They love clover, basswood, sumac, black locust, dandelion, apple blossom, burdock, and goldenrod. The list goes on and on.

Hive Types

There are three common types of beehives: the Langstroth beehive, the top bar beehive, and the Warre beehive.

The Langstroth hive is the standard hive used in most developed countries, such as the United States. Its appearance resembles that of a filing cabinet, but do not let that fool you. The top is removable so there is a lot of easy access, and the entrances for the bees are small slits between the body and the bottom baseboard of the hive. Inside the body of the hive, there are vertical shelves and a thin layer

Honey extractors are standard built for Langstroth hives, making it easier during harvest time.

of hexagonal wax sprouting from each shelf, which is a guideline for the bees when they begin making their combs. To expand the hive, it is easy to just add another super (a piece of the hive body, but thinner) to the top. First you have to remove the removable top, add the super, then just place the top on the added super.

Langstroth hives are simple to sustain and fix, and harvesting honey from them is a breeze. To collect the golden honey, just slice off the tops of the combs, place in a honey extractor, and spin! After extraction, you can effortlessly place the shelves back inside the hive. This is nice for the bees because they won't need to start from scratch in the new season, and is nice for you because that means you will get more honey!

However, with every good thing, there is a downside. With a Langstroth hive, you cannot know where the wax and guideline layers begin, and the wax guideline combs are usually larger than the size of natural wax combs. This can make it difficult to detect problems such

A top bar hive

as varroa mites, which pass on a disease called varoosis, which weakens the bees and can kill an entire colony. Langstroth hives also make it hard to know what a colony has been exposed to.

A top bar hive is a long wooden structure resembling a covered bridge but with a v-shaped bottom half. The pitched roof keeps rain and detritus out while the magic happens inside. Remove the roof, and you find a flat surface made up of the horizontal slats that serve as the upper edge of the frames. These slats overlap the larger structure of the hive, therefore holding the vertical honey frames in place as they hang down inside the hive. The v-shaped vertical frames nest side-by-side and have angled ridges so the bees have room to build their combs on the surface. Bee entrances are located in the front or on the sides in the forms of slits and/or holes. This type of hive is a great DIY project, and if you want to get fancy, you can include an observation window to watch your bees make their combs. With a top bar hive, you don't need to buy or build supers, so it is very inexpensive. If you plan on making other products besides honey and plan on using the wax, you can guarantee the products are pure because everything comes from that one bee colony.

A Warre hive

If your hive grows to maximum capacity in one top bar hive, it is easy to split the colony and move one half to a new hive. Also, during the winter months, you have to make sure the bees have food and water. Bees need at least twelve flats of honey to get through the winter (this is about 70 pounds of honey), and for water, the condensation caused by breeding suffices as their water supply.

Note: Because the bars are removable, if a bee builds one comb on two bars it will be ruined, so you must make more frequent inspections (at least twice a week). You also have to make your own equipment like feeders, queen excluders, entrance reducers, and beetle traps because each top bar hive is different. This type of hive yields less honey than a Langstroth but is useful for raising queens and larvae.

ROBBIE MINER

Robbie Miner is from southern Vermont, a place that fostered in her, from an early age, an appreciation for the world of plants and insects. Since 2008, Robbie has lived in America's West searching out new adventures. She lived in northern New Mexico and interned under top bar master beekeeper Les Crowder, and worked for many organic farms. A wanderer by trade, she works seasonally, trying to soak up the wonders and teachings of this earth. A self-taught herbalist and plant forager, she spends her off-time hiking and climbing high country. In the winter, she ski patrols and now lives in the Central Cascades of Washington.

Top bar hive materials:
- Any type of wood that hasn't been treated with chemicals. If it is, the honey will be impregnated with the chemicals.

- White paint helps control temperature.
- Bees need twelve flats of honey to get through the winter and regular hives have twenty-four.

If bees swarm:
- Put honey, wax, and lemongrass in a box. Usually they will congregate in the box.

What is permaculture to you?
Everything works together. Bees pollinate plants and we reap the honey from the pollen they collected from the plants we tended to.

What types of flowers do bees like?
Herbs, fruits, wild flowers like clovers, alfalfa, goldenrod, squash blossoms, bean flowers, marigold. Bees don't like genetically changed plants like tulips or lilacs.

What is your advice for treating sick bees?
Put juniper in smoke and puff the bees with the smoke. It will cause the mites to fall off the bees, decreasing the number, allowing the bees to clean the rest out themselves. Don't use antibiotics when bees are sick, it kills the good bacteria in their stomach too. Requeen.

The third most common beehive is the Warre hive. If the Langstroth and top bar hives made a baby, the result would be a Warre hive. This hive allows for little intervention and is a great option for beekeepers interested largely in garden pollination. From the outside, a Warre hive looks like a little tower. Internally, it is structured similar to a top bar with its hanging frames; however, these frames hang within boxes that stack on top of each other, upright like a Langstroth. This gives an impression of a tree hollow. There is a box at the top where wood shavings are placed between two layers of cotton called quilts. This little box is created to regulate temperature and moisture levels. The bee entrance is at the front's bottom and you never need an entrance reducer because it is narrow enough.

Unlike a top bar but like a Langstroth, you do use supers, but instead of adding them to the top, you add them to the bottom. This allows bees in a Warre hive to build their combs upwards. There is low maintenance involved, and the hive is pretty much self-regulating.

A Warre hive can be expensive if you were to buy a ready-to-go one, and adding on supers is difficult if you were to add them by yourself (it is easier if you have someone help you). Because the entrance is so narrow, you can't use an entrance feeder, meaning you will have to make a top one or place a feeder in a super at the bottom, but it becomes annoying to refill. Feeders are devices that allow you to feed your hive sugar syrup when they are low on or out of nectar. Feeders come in many different sizes and shapes and are customizable based on your hive type.

Bee Equipment

Beekeeping can be as lavish and expensive or as low-key and independent as you want. However, there are some things you need to spend money on, and that is the equipment needed to stabilize your hive.

For beginner beekeepers, there are nine essential pieces of equipment needed:

1. Hive tool
2. Frame lifter

3. The Italian
4. Helmet with veil
5. Gloves
6. Full body suit or jacket
7. Smoker (necessary at first; optional later)
8. Bee Feeder
9. Brush

The first thing you will need every time you go to inspect your hive is a hive tool. Now, there are a few different ones, but the best thing to do is start off with the most common one and once you get the hang of things, try out a different tool and see how you like it. The beginner—and most commonly used—version is a metal tool with half of its 7–10 inch body painted and the other half left uncoated. It has one curved end that is used for scraping and prying while the flat end is mainly for prying things apart. Also, in the middle, there is usually a slot made for pulling nails free. The second version of the hive tool is the frame lifter, which is 9–14 inches long and is made for—wait for it—lifting frames. One side is made mainly for that purpose, with an offset hook, and the other side is made for scraping and prying. The last honey hive tool is called the Italian, which is 12½ inches long and very thin in its width. It works best for cutting honeycombs off the sides of top bar hives.

Next, you always need a helmet with a veil. Sure, maybe some professional beekeepers don't use one because they are fearless, but it can become really scary with a multitude of bees flying around you and nothing covering your face and neck areas. Even if you don't like wearing a full body suit, it is best if you still wear a helmet with a veil.

Most people are not immune to bee stings, and that is why you should have a pair of gloves. Now, they aren't impenetrable, meaning you can get stung wearing just one pair so it helps to double the dose and wear two pairs of durable leather gloves or other gloves made of strong material. Doing this will immensely decrease your chances of being stung through your gloves.

For beginners, it is recommended to have a full suit, but if money is tight, a jacket is better than nothing. If you have had experience with bees, a jacket probably isn't something you need, but it is sometimes nice to have when the bees are especially irritable.

Some beekeepers don't believe in using a smoker, and you can be the judge of whether or not you believe in using one, but it doesn't hurt to try it out when needed. Beginner beekeepers should try to make it a mandatory buy and judge it later. It is helpful because it subdues the bees by hindering their communication and causes them to believe there is a fire, so they become distracted by mercilessly consuming honey.

In order to keep your hive alive, you need a bee feeder. The type of feeder you use depends on the type of hive you use, and you can buy, order, or make them. A top bar hive will require you to make your own custom feeder, but for Warre and Langstroth hives, it is easy to buy one.

A tool you should never be caught without is a brush. It is mandatory when visiting the hive during inspections and is usually just a wooden handle and back with flexible brushes that can easily push bees out of the way. It looks like the brush that comes with a dustpan and nests in its handle. Obviously, it is for gently sweeping bees off surfaces, but the bees soon begin to hate it, so make sure you are properly prepared for being attacked. Gloves and a brush go hand in hand.

GORDON FISHER

Gordon has over 100 acres with his family in West Brattleboro, Vermont. He has a small sugaring operation and a 20-acre wood lot. He has a raspberry crop, and boils sap for maple syrup. These days, he is partially retired from heavy lifting, but he still keeps bees.

So how long have you been keeping bees?
I've been a beekeeper for over forty years. I'm seventy-eight.

Can you tell me about your bees?
I've had some loss, in the winter, but mostly from mites in the last few winters, but all in all, I've had pretty good luck with them.

I have two hives that are homemade with a standard Langstroth body. One hive I started in the spring after someone gave it to me—a four-way split—and the bees have done very well.

Usually, if you get one super of honey the first year off a full swarm, you've done well. A lot of years you don't get anything. This year I've taken one off of one hive, and, on the other hive, I've taken three off already, so it's been a good year this year so far.

In the winter, I usually have some die-off, but mostly due to the mites. I try to tend to them and keep the bees in good shape. I've kept bees for almost forty-five years!

I eat a lot of honey—probably more than I should! It doesn't bother me, and I feel I'm pretty healthy for doing so.

Gordon, since I am going to be helping you in exchange for learning the ropes, what are some tips for other beekeepers?
Well, I tie a few black plastic bags to the electric fence that surrounds my hives. The bees don't like anything dark colored, so

they see this waving and they know it doesn't hurt them. I've had less trouble with the bees since I put the bags up.

When you go in, do you suit up?
I just wear a veil. I put the veil over my head and tuck it inside my shirt, under the collar. I also wear gloves—sometimes, not all the time—and I rarely get stung. So, my theory is it's how you handle the bees. I do use a little smoke, however, when I first start going in to work on the hives. But just a little bit.

Another tip I have for other beekeepers involves using smoke. If you use a little Staghorn (non-poisonous) sumac with the smoke, it helps get rid of the mites. The mites don't like it. That's what I've been doing, instead of treating the bees with chemicals. I have a very healthy hive; at least they seem to be!

Tell me about bee culture.
The worker bees, the females, bring in nectar and pollen. They are busy all the time. The drones are still in there, but the rest of the hive will be kicking them out, probably next month in October. The males die then, but there might be a few that make it over the winter. In the spring, the queen will lay unfertilized eggs, which are the drones. Any egg that is not over twenty-four or forty-eight hours old, could become a queen, or a female worker bee.

How long have you had your queen?
You should only keep them for one year, and then you should "re-queen" them. You either destroy the old queen, or let the hive raise one early enough in the fall or summer. Or, you can buy a queen and introduce her into the hive that way. I've done both ways, and I've even sent away to California for them.

Can you tell us more about bee culture and harvesting honey?
Bees are good for pollination, and without the honeybees, you would probably only have around eighty percent of a crop. Bees are very sociable with each other, but not with other hives. What I do in the fall, is take the honey off, and put it in an extractor. I cut the caps off —you can use an electric knife, or a hot water knife for this—then you put it in a spinner—and it spins the honey out—by motor, or by hand. I have used the hand spinner most of my life. After that, you strain it and jar it up. It's all-natural.

Does it take a long time, with the hand method using the spinner?
Not too bad. It is work, but it's worth it!

How many bees do you have?
It is hard to say, but it could be close to 150,000 in both hives. I think the further hive has more. The queen will lay up to 2,000 eggs per day in the summer.

(Gordon cautioned me to not get in the flight line of the bees when I was taking photos.)

What do you do with the honey?
I give a lot of it away, to family and for presents, but I also sell a little bit of it.

What do you do to keep the hives safe from predators?
You get bears, and I've had that happen to me several times. If you don't put up an electric fence around the area you will have

problems, and you will be apt to lose the hives no matter where you are. Over the years, I had several incidents when bears came through. I put up the electric fence, and haven't had any problems since. If the bears come up to the fence, they get a pretty good shock, and since they are down on all fours, it's more of a shock than for us.

Any other tips to tell us about the "Gordon Fisher method"?
In the winter, I have a system with 2 x 4's nailed together with a hole up the middle and a screen over the top so the mice can't get through. I fill the little body full of wood shavings, and that protects the bees. I also put tarpaper over the hives, and staple it on. That seems to protect them. It seems to work. However, if you don't take care of the mites, you are in trouble.

How do you find out if your colony has mites? Do you just examine the bees visually?
It is hard for me because I can't see the way I used to, but the mites are hard to see anyway because they are so small, around the size of a pinhead. But, they have different screens you can put on the bottom of the hive, so when the bees walk across them, the mites will fall down through the screen, and they don't get back on the bees. I think the smoke treatment with Staghorn sumac is one of the best ways to get rid of mites. They drop off, and go through the screen and can't get back up.

Are you concerned about Colony Collapse Disorder (CCD)?
I am not sure, and I've read different studies on it, but scientists are not sure yet. They said we have to have a 5–10 year study in order to find out why it happened. First, they thought it was the cellular towers, and then there are other things that they are working on. A byproduct that they use to spray tobacco and other commercial vegetables was used, and they think CCD could be coming from that. They will study it, and we'll find out.

What about using chemicals?

Well, I think chemicals play a part in CCD, certainly, but also the mites are a big part of the problem. It's all I can think about.

Do you use any chemicals?

No, I rely on the smoke. I have never given my bees any fallow brood medicine. I have had very good luck. If you have very strong hives, they say that it will cover up the spore that causes that—they will wax it over and survive. If you use chemicals, it would keep breaking the bees down to where it wouldn't be any good for them anyway.

So you are an advocate for organic beekeeping?

Yes!

(After the interview by Gordon's apiary, we walked down to his raspberry patches and his peach orchard. He told me he started growing peach trees from burying the pit from a local orchard in the ground.)

Gordon, can you tell us about your raspberry crop and why the bees are good for it?

Well, I've had a very good year this year with the raspberries. I have never had to spray my raspberries. If I needed to, I would only use something organic. The bees are right over there, and the raspberries are down here, so it makes for a short trip. The bumblebees do a lot, too, but there are so many honeybees, and they are

so close, that it's convenient. It works for me. The bees gather the nectar and pollen, which the bees use for their source for food.

Does planting flowers help the bees?
Yes, they say that it doesn't pay to plant for bees, but I disagree. If you didn't have anything there nearby, then what would they get? If you plant something—and everybody does it—even flowers, or a small crop like I do, it will help all kinds of bees.

I'm thinking of keeping bees. What advice do you have for me?
The first thing is you should have an electric fence to keep predators, like bears and coons, away. Skunks will even get in there and stir up the bees and make them very unmanageable. You should have two big bodies (Langstroth), with ten frames in each one and a couple of small (spare) supers. And you have to have queen excluders, a bottom board, and inner and outer cover. There is a lot to it. I love it because I've been in it so long. When you make your order in January or so, I'll help you. I can tell you a lot of things about keeping bees, but I still don't have all the answers; nobody does. What one hive will do, the other one might not. It is good to have 1–2 hives to start with, to get the feel of them. My advice is start small, and see if you like them and see how you make out the first year or two.

How does it work getting the honey off?
The bodies will weigh approximately 35–40 pounds apiece. I can't lift them off the way I used to. I use an escape board underneath and the bees will go down and won't come back up. Then I can take the honey off without the bees there. Otherwise you have to take one frame at a time and shake the bees off, which is a lot harder and takes longer.

I have ten-frame hives, but in mine, I leave nine to make a comb. The bees will draw it out further, and you'll get more honey in the long run—you will also have less frames to contend with—that is why I only put nine in there, and space them out. I learned a lot from the old timers. They were from around here and when I was a young man, I just started talking with them.

Can you tell us how you got started as a young man?
I found a bee tree one time, and we ended up cutting that down and taking the bees out. I actually made them survive the first winter. That was pretty good and a lot of fun, but it was scary at times. Now it doesn't even bother me one bit. I found that bee tree out in the woods. You can still harvest honey in the wild, but today—with the mites—it is a lot harder to find a bee tree than it would be years ago. There are a lot of things we don't know about bees. Every hive is different. People say that I am crazy, and they are afraid of bees, but I love my bees because they do so much good for crops and different things.

Helpful Equipment

Besides the bare minimum, there are also a few tools and equipment that come in handy when taking care of your bees. Entrance reducers are something that help keep the bees warm during the winter months and keep pests like varroa mites out. The only hive that doesn't need an entrance reducer is a Warre hive because the entrances are narrow enough.

Speaking of varroa mites, it may be smart to buy a varroa screen, which creates a safer environment for the bees. It is always better to be safe than sorry, but if budget is an issue, the hive can live without it.

A hive stand is something that helps in a few different ways. You can either buy one to your liking or build one to custom fit your hive and location. The stand keeps the hive off the ground, which keeps the hive safer from flooding and being stepped on. It is also easier to create some sort of slope needed for the hive to thrive.

10-FRAME LANGSTROTH BEEHIVE
CONSTRUCTION DETAILS FOR 3/4" THICK LUMBER

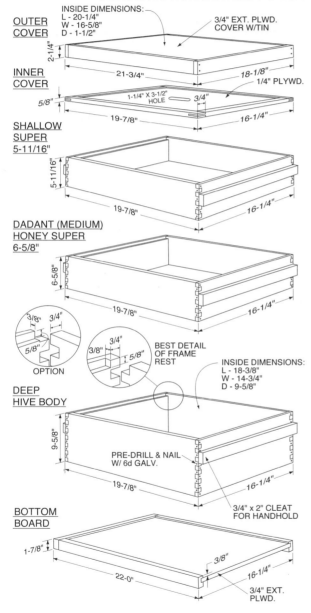

OUTER COVER
INSIDE DIMENSIONS:
L - 20-1/4"
W - 16-5/8"
D - 1-1/2"
2-1/4"
3/4" EXT. PLWD. COVER W/TIN
21-3/4"
18-1/8"

INNER COVER
1/4" PLYWD.
5/8"
1-1/4" X 3-1/2" HOLE
3/4"
19-7/8"
16-1/4"

SHALLOW SUPER
5-11/16"
19-7/8"
16-1/4"

DADANT (MEDIUM) HONEY SUPER
6-5/8"
19-7/8"
16-1/4"

3/8" 3/4"
5/8"
OPTION

3/4"
3/8" 5/8"
BEST DETAIL OF FRAME REST

INSIDE DIMENSIONS:
L - 18-3/8"
W - 14-3/4"
D - 9-5/8"

DEEP HIVE BODY
9-5/8"
PRE-DRILL & NAIL W/ 6d GALV.
19-7/8"
16-1/4"
3/4" x 2" CLEAT FOR HANDHOLD

BOTTOM BOARD
1-7/8"
3/8"
22-0"
16-1/4"
3/4" EXT. PLWD.

The species of wood used to make a beehive can vary depending upon what is available in your area. The minimum thickness should not be less than 3/4". If you are using standard dimensional lumber, you can use 1x8 (3/4" x 7-1/4") for both shallow and medium super, and 1x12 (3/4" x 11-1/4") for the deep hive body.

Start by cutting the boards to length. For fronts and backs, cut them a smidgen over 16-1/4". For sides, cut a smidgen over 19-7/8". Cut to exact size before assembling. At this point, cut box joints on all the board ends. Rabbet joints are an acceptable alternative.

Now that you have the joint cut and the boards cut to finished size, cut the 5/8" x 3/8" rabbet on the 16-1/4" boards stopping just short of the box joint pin at each end. (Chisel these square after the boards are assembled). Note detail of frame rest at left. Pre-drill holes for nails in each pin.

Assemble boxes with glue and nail each pin with a 6d galv. nail. Attach 1x2 handholds with screws and glue. Attach metal rabbets on the frame rest notch. Fill any holes and paint all exterior surfaces, both top and bottom edges, with primer and finish top coat.

BILL MARES

Bill Mares is from the Vermont Beekeepers Association. Raised in Texas, educated at Harvard, Bill Mares has been a journalist, high school teacher, and member of the Vermont House of Representatives. He has authored and co-authored fourteen books on subjects ranging from the Marine  *Corps to workplace democracy to desert travel to presidential fishing. His hobbies include running, beekeeping, singing, and fly fishing. He lives in Burlington, Vermont with his wife of forty-three years, Chris Hadsel. They have two sons.*

How long have you been keeping bees?
Forty years.

What inspired you to begin as a beekeeper?
My wife and I lived in St. Johnsberry, Vermont where I had a friend with a farm who loved beekeeping. Selling bee products and teaching younger people about beekeeping created such a joy in my life that I have continued doing it through the years.

Any tips for beginners?
Patience. Don't get discouraged. For the first year work with an experienced beekeeper until you get a feel for the trade. Bees need and deserve a lot of attention and they aren't like any other animal. Spend time doing it—that's the only way you will get better at it.

Tips for experienced beekeepers?
Keep up with pests and viruses. You can never know all there is to know about bees. Make sure to keep up with other beekeepers.

What products do you make?
Just honey—eight-ounce Ross Rounds right out of the hive.

How many hives do you recommend someone begin with?
Start with two because sometimes hives die. If that happens, you can then split the living hive in half. If you have just one hive and it dies it can be a big psychological downer. Also, it is like having two children—they are each different and you can learn different things from both.

What beehive do you use/like?
Langstroth.

Where do you get your bees?
Don't have to buy bees anymore, just new queens and then split the hives. I get them from a bee supplier in Greenwich, Massachusetts.

Where is a good spot to put a beehive?
Ideally have it face to the southeast so it catches the first sun. Create a windbreak.

How do you attract bees?
I collect swarms that leave a hive. However, it is a lost practice because swarms usually don't last the winter.

What do you do when bees or the whole hive becomes sick?
Depends on timing. Most of the time, I change the queen and see if anything changes. In the winter, dysentery occurs but they can fix it themselves by spring. With varroa mites, just try to keep the level down as much as possible.

What flowers do you find bees are most attracted to?
It depends on the season. Goldenrod is great in the fall and blueberry flowers in the summer months.

Urban Beekeeping

"The single greatest lesson the garden teaches is that our relationship to the planet need not be zero-sum, and that as long as the sun still shines and people still can plan and plant, think and do, we can, if we bother to try, find ways to provide for ourselves without diminishing the world."
  —Michael Pollan, author of *In Defense of Food* and *The Omnivore's Dilemma*

How to Get Started

Backyard beekeeping is just that, even when your backyard is an alleyway and the only means of creating an apiary is using a rooftop. It seems a bit far-fetched, maybe, but with the correct knowledge and motivation, it is possible. There are, however, a few things you need to take into consideration when making a home for bees where many other people have already made their homes.

If you have a plausible backyard for keeping a hive, or maybe a few, something you should have is a fence. Like other domesticated animals, which is something a bee is and is not, some people do not appreciate them flitting in between their ankles and through their line of sight. A fence allows for avoiding such incidents. The height of the fence—ideally six feet—forces the bees on a higher flight path, rather than straight forward. If you don't want to build a solid fence, you could opt to create a barrier out of tall shrubs. The fence or shrub barrier also makes for a good wind-breaker. Because people will not see the bees as much as they would without a barrier, they may even forget that you have hives.

Swarming is something you must pay attention to. The colonies that swarm are usually the ones with an abundance of strength and an adequate queen. Often, if the hive has a young queen, they do not tend to swarm as much or at all, but they are still healthy. If swarming is becoming a problem, try to switch out queens every year. If that is something you do not want to do, another option for limiting swarming is to place a bait hive ten feet to thirty feet from their main hive. Optimal bait hives are made out of old hive bodies or have an old frame. This is because bees enjoy places where they themselves or other bees have lived before. These bait hives should also be elevated and placed in total sunlight. If by some chance the bees still swarm and you catch them in time, take with you a box filled with some wax, honey, and lemon grass, recommends Minor. The lemon grass smells like the queen, and once you gently push a few bees into the box, word will spread, and the other bees will follow.

A stung civilian is not a happy civilian. Remember this when working in the hive. Fortunately for you, there are ways of preventing a mass of angry bees. Mostly, timing will be your saving grace—that and weather. If it is less than ninety-five degrees Fahrenheit outside, the bees won't become as agitated. Try and do the work needed to be done when the hive is fully consumed by sunlight, this way, the field bees will be out collecting nectar, leaving less bees to upset in and around the hive. You should also know how many people are out and about. This does not necessarily mean to go out into the streets and count them; just be aware of any gatherings happening in the area or if a lot of people are traveling by your yard. Be a courteous neighbor as well as a saavy beekeeper.

JOHN HAYNES

John Haynes lives and keeps bees in the backyard of a suburban area in Novi, Michigan. He and his wife started taking beekeeping classes as an experimental hobby for future retirement activities and quickly became hooked. John, currently an engineer both with his bees and as his professional career, has come a long way from a take-home hive from his first class to owning five successful backyard hives from which he makes products such as tinctures and candles. He is now on his second year of being chairman of the Southeastern Beekeepers Association and a wonderfully studied bee enthusiast.

How long have you been keeping bees?
I've been keeping bees for five years.

What inspired you to begin as a beekeeper?
Well, I was just looking at it as kind of a long-term hobby for retirement. I saw some publicity then that the bees were dying; they were in the news quite a bit. So I thought maybe we can invite some bees in. When I started my class, they required that you raise a hive as part of the class, and at the end, you had to take the hive home. About twenty-six people were in the class; my wife took it a couple years later. I took an advanced beekeeping class afterwards at the Southeastern Beekeepers Association.

Any tips for beginners?
I would absolutely recommend taking a class. I don't think I would've been able to do this without a class. As soon as I signed

up, I started reading. The reading was helpful, but self-education is really critical. A lot of things change almost every year; they've been changing within the last 15–20 years, actually. Man, ever since tracheal mites and viruses, parasites make the bees more susceptible. You have to be aware of all the things that are happening in and around your hive. It's not like the old timers, where they had to set them up with a box in a field and had more bees than they knew what to do with. Find a mentor that you can go to with questions or problems. I've been a chairman on the beekeeping association for the last two years, and we have many classes for different experience levels.

Please relate a memorable experience you had as a beekeeper—humorous, instructional, tragic, or all of the above.
Well, one is sad, and one is not so sad. The first year that I started, we had one hive with our class, and that died over the winter. That was a time when state-wide bee mortality was about 50 percent; it was pretty demoralizing. We felt bad that we had all these bees that are now dead. Really, we just had a terrible, guilty feeling that all the bees died, so we learned from our mistakes.

A pretty exciting thing was when our bees swarmed a couple times, and swarming is very exciting. Once, we were in the backyard when our bees decide to swarm. They make this big cone of bees and go land somewhere. One of the multiple swarms landed in a tree next to their swarm. Last year one hive swarmed, and I just watched it fly away. When they swarm, they go out to look for a new home. Our bees come from Georgia, and if that hive swarms the first year, they're probably going to die. It's the multiplying trait of bees; it's how they expand the species. If you have a hive that gets too small for them, they gotta split. So, a half or two thirds of them will leave to save the existing colony and start a new one elsewhere. It's how they keep the species alive. It's their instinct to swarm, and as beekeepers, our number one objective is to keep them healthy, and to manage populations to prevent them from swarming.

What products do you make?

I make candles for gifts that I give away at Christmas. I also make a propolis tincture, which is an antibiotic kind of an antiseptic material that they create inside the hive. It's used to fill up cracks to keep the hives clean. You can collect it in the hive, and they have a tendency to fill said empty crevices with propolis. You can also put things in there that they will fill up with propolis. It's a liquid that you mix with 190 proof alcohol, and then you can add it to tea, or use it as a topical application (Author's note: propolis has a long history of medicinal use in fighting infections and assisting healing). The biggest bee product today, other than honey, is bees themselves. Those in and of themselves are a product—raising clean nucleus hives and selling those. If I have good survival, I would sell them. It's for the hobbyist and for commercial people. A nucleus hive of local bees are very valuable.

How many hives do you recommend someone begin with?

We started with one, which was a bad idea; I would say two. Even for the first year, as you get more experienced and want to do a split level hive, you might figure that half of your hives are going to die, so you want to raise more to replace them. If you have two hives and you want to have a nucleus hive, you'll want at least three into winter. Two that you want to survive, and one as a replacement.

What beehive do you use/like?

We don't use top bar hives; we use langstroth hives, which are the standard length hive bodies. They're interchangeable, if you want to go to any catalog. I make all the wooden stuff, and I actually use plastic frames because they're durable; they don't have to mess with the wax. They don't absorb viruses or anything like wood would. I have good luck with plastic.

Where do you keep your bees?

We keep them in the backyard. We have a regular subdivision lot: a quarter of an acre. We have five hives now and about five nucleus hives. We want to keep it small enough to where they're not flying into the neighbors' yards too much.

Where is a good spot to put a beehive?
I think we're pretty lucky because we put in a small pond before beekeeping, and it turns out that it's a great water source. If you're doing urban beekeeping, you don't want your bees going over to the pool next door. They love the pond, and they don't have to go very far to get their water.

What flowers do you find bees are most attracted to?
Whatever you can plant in your yard is not going to feed even one hive. They really like mint; it does contribute a nice minty flavor to the honey. Lavender, as well. The most beneficial florals are trees. Black locust, crabapples, the right kind of basswood. Those are where you get the biggest bang for the buck; a flowering tree is going to take care of one hive's needs. If you have bees out in the middle of farm country, there's not really much out there unless there's some good stands of trees with good floral production.

How to Work with Your Backyard Bees

"When the sun weeps a second time, and lets fall water from his eyes, it is changed into working bees; they work in the flowers of each kind, and honey and wax are produced instead of water."

—from ancient Egypt

Obtaining Bees

FREE BEES
If you don't want to spend money and want a budget saver when starting your own apiary, capturing bees for your hive is the way to go. Out in the

wild, bees usually form swarms and divide into separate colonies for a couple of reasons: when the queen falls ill, when the queen is injured, or when they have reproduced too much and have simply outgrown their original hive. Collecting a swarm, with the proper equipment (such as a beekeeper suit, container, cardboard and/or smoker), is actually easier than you think. Bees from the wild are often compliant; although, there can be a few that are ready to strike, so be prepared. If the hive is on the limb of a tree, carefully cut the limb down, and gently shake the hive's contents into a container. If the bees have taken up residence on a flat surface like a fence post, you can softly brush the bees into a container with a piece of cardboard. If you don't want to use cardboard, blowing a small amount of smoke at the bees to direct them into the container works, too. However, wild bees are more prone to carrying diseases or liable to have weak DNA, which makes for a weak honey supply.

Often times, the queen is difficult to find, so you may have to buy one from a local beekeeper. Check your state laws before going bee-hunting behind your house because some states have laws against capturing bees.

If you are a beginner beekeeper, buying bees may be easier for you. To order some, just contact a local beekeeper. If you can't find one, you can also order bees off the internet. In most standard packages, there

is a queen, worker bees, and a feeder filled with food. But because the queen bee in the package usually isn't the workers' queen, they have to be introduced. There are two ways to do this.

The direct method involves the queen being released immediately with the other bees. Because they aren't used to her, the queen will most likely be attacked by the worker bees. However, if she is not attacked and they accept her, the hive begins reproducing and honey production quickly commences. Thus, the hive works more efficiently given such short notice.

The other is an indirect method that allows the worker bees to become familiarized with the queen as they eat through food in order to get to her. This is safer because, chances are, they will not attack her like they would with the direct approach. Albeit, this method takes longer for the hive to commence work.

If all of that seems like too much work to you, there is another option called a nucleus hive. They are prepared colonies, with their own familiar queen, ready to start honey production. There is a common faulty aspect involving a nucleus hive, however. Sometimes, the queen may be weak or old, making for an unsteady colony.

MICKEY ROWLAND

From cozy Nantucket cottages to multiple structure compounds, Milton "Mickey" Rowland designs living spaces that enhance the lifestyles of his clients at Milton Rowland and Associates on Nantucket Island. He also keeps bees on the island of Nantucket. Originally, Mickey wanted to keep bees in order to help pollinate his fruit trees on his seaside property. His super equipment (hive boxes and frames) started out small, and now he has over ten hives and produces enough honey to sell on the island and give out as presents. Rowland has learned how to keep bees and maintain his hive, even producing different grades of honey: light, medium and goldenrod, that follow the seasons and offer a taste of the nectar from the island.

How long have you been keeping bees?
About twelve years.

What inspired you to begin as a beekeeper?

I had about fifteen fruit trees in my back yard that weren't producing very well, so I thought that getting bees might improve the pollination.

Any tips for beginners?

Bees are pretty capable of taking care of themselves, but they do need inspecting on a fairly regular basis. The biggest concerns are swarming and mites. Open the hive and inspect the bees every couple of weeks—maybe even every week in the late spring and early summer. Look for signs of overcrowding which will always lead to swarming. This is very undesirable as you end up losing half of your bees plus the queen. It can take a few weeks for the new queen to start laying eggs, so you have lost some prime population building at the most productive time of the year. If the first hive box is full, add a second. If that is full, start adding honey supers if you haven't already. If it's really crowded, you may need to split the hive. Check for mites throughout the summer and fall, and treat with an organic treatment if needed.

Any tips for experienced beekeepers?

I have a few different experiments going each summer, and I learn a little from each one. My only advice to an experienced beekeeper would be to try something new in one or two hives each year. You may not make too many changes to your systems, but it will always add to your general knowledge base, and that is how we slowly improve the processes over the years.

Please relate a memorable experience you had as a beekeeper—humorous, instructional, tragic, or all of the above.

One summer, I discovered that one of my hives had just swarmed and landed in a nearby tree. Collecting a swarm, especially one that is about fifteen feet up in a tree, is always a unique adventure, and this one was difficult but inevitably successful. I considered this very fortunate to have recovered a swarm from one of my own hives and thought the logical move would be to put it back where it came from. After all, that hive had just become my weakest hive, and at the time, I didn't have any additional hive setups to put the newly captured swarm in. I carefully combed through the original hive and removed all of the newly formed queen cells. This was of course necessary to avoid hatching a new queen who might possibly kill off the original productive queen, resulting in the loss of at least a couple weeks of egg laying in the middle of the summer. Having completed my task, I poured the captured swarm back into the hive box from which it came. Well, a hive that has just swarmed generally becomes a pretty small and docile colony of bees. That changed dramatically the instant that I returned the vagrant bees to their old home. It was as if

a small bomb exploded within the hive. The bees came pouring back out in all directions with a sound and volume I had yet to hear from a bee hive. Most of them seemed to realize that I was the cause of their recent unexpected return home and attacked me with a vengeance. Fortunately, I was fully suited up with veil and gloves, but being in the middle of the hot summer and with the added stress of the situation, my thin bee suit clung tightly to my sweaty arms and shoulders. I'm sure I have never been stung as many times as I was that afternoon. Needless to say, nothing about that plan worked out as I hoped. I'm sure that all of the returned swarm eventually departed (for the second time), and the hive remained a very "hot" hive for the rest of the summer.

What products do you make?
Mainly honey. Over the years, we have made some candles, but it is a lot of work and always very messy. One year, Robin made a dozen or so tubes of lip balm. That went pretty well but, for

some reason, hasn't been repeated. And one year, we made fire starters by setting dry pine cones in small paper muffin cups filled with melted wax—pretty successful but also a one-time event. Producing honey is a lot of work, but the reward is always clear and nearly guaranteed. The others are less consistent. I still save bricks of wax, thinking that someday I will find the time to use them in a productive way.

How many hives do you recommend someone begin with?
It seems that most of us start with just one hive, probably because the equipment isn't cheap, and the uncertainty of how long the fascination will last is unknown at the beginning. I would recommend two hives if possible. That way, it is easy to gauge whether one is doing poorly or really well. And if you lose a queen for some reason in the middle of the summer, you can always make a new queen using frames from the other hive.

What beehive do you use/like?
I have always used the Langstroth type of hive. Deep boxes with ten frames each, and medium-sized honey supers on top.

Where do you get your bees?
I used to be able to have them sent from somewhere in the South through the post office, but they don't do that anymore. Recently we have collectively ordered packages, driven up to northern Massachusetts, and brought them back to the island in a covered pickup.

Where is a good spot to put a beehive?
I used to keep my hives along a cleared path in the pine woods below my house. It was a beautiful spot, and the shade from a few trees made working the bees at midday rather pleasant. I started experiencing symptoms of stress in the hives which I believed was caused by cool damp conditions. I have since moved my hives to an open field where they get full sun all day. I believe they have benefitted from the move.

What flowers do you find bees are most attracted to?
I have never tried to attract bees. I believe some try to attract swarms with pheromones or other such stuff, but I never felt that would be productive. Bees will almost always return to their hive. Sometimes they will end up in a hive placed right next to their home, but trying to attract stray bees isn't done (as far as I know).

There are many reasons that a hive starts to decline. And that is why it is important to check on your hives regularly. This way, you will see if there is a drop in egg laying or brood coverage in the frames or unusual looking cells. Covering this subject would take up several pages. When I suspect a problem, I first check to see if the queen is present and looks healthy, and then I look for mites inside the capped drone cells or an unusual number of dead bees outside the hive. The remedies vary but may include re-queening (a drastic move) or adding frames from other, strong hives to increase the strength and population. Sometimes it is best to just let it go and hope the other hives don't succumb to the same problem.

I've never figured this one out. I don't actually see a lot of honeybees on my fruit trees—more wasps and bumblebees than honeybees. They all seem to fly up out of their hives and disappear into the sky. I have a large fresh water pond nearby and the biggest agricultural farm on the island within their range, so they could be going anywhere. Plants on the island that I have noticed attract honeybees include privet, knotweed, goldenrod, dandelions—the usual stuff you probably see in your neighborhood.

Hive Maladies

There are many hive maladies, and each has their own cure. When it comes to something that cannot be pinpointed, Mares states that, often, changing the queen helps if the disease or illness was indeed caused by her. "It all depends on timing," says Mares. Dysentery is an illness that occurs during the harsh winter months, but according to Mares, if it is not severe, the bees usually get better by spring time.

Varroa mites—the honeybees' nemeses—are something that usually cannot be prevented unless you have a varroa mite screen. Even then, they can still slip by. As previously stated in our hive types section (page 21), these small creatures pass on varroosis, which is fatal to the bees who contract it and possibly the entire colony. The most logical way to deal with this, Mares says, is to keep the population level down. Beehives are "much more demanding of your time and attention," claims Mares. If you have access to juniper, putting it in your smoker also gets rid of the bees. According to Minor, puffing the bees with the juniper smoke causes the mites to fall off the bees. Of course, not all mites will be gone, but the level will largely decrease. For all natural beekeeping, Minor also advises all beekeepers not to use antibiotics on their sick bees. The medicine kills both the bad and good bacteria in the bees' digestive tracts.

Organic Gardening Techniques

Some people may say that using certain types of chemicals, such as ones with organic bases, is okay. But when trying to create a healthy ecosystem for bees, chemicals are never the way to go. Organic gardening techniques may be more time-consuming (though only by a margin), yet it makes for happy bees. Pesticides are meant to keep harmful pests out, but they harm the good ones, like bees, too.

Sometimes, however, those who have an intent to go organic do not always meet their goal. Deceit from large gardening stores is usually what pulls them out of commission. Plants sold in such stores claim to be bee-friendly, but the pesticides and neonaticides they have been treated with tell a different story. The aspect that makes it especially frustrating is that the treated plants are never labeled, leaving it up to chance. The best thing you can do to avoid accidentally buying toxic plants is to buy your plants and seeds from a local grower and seller where pesticides are not used. And because bees can fly up to three miles away to gather food, you should definitely be aware of any gardens or fields that use pesticides in your area.

If garden pests are the problem, there are non-toxic pest controls. The most natural technique is using predator versus prey to your advantage. Plant crops that will benefit you in yielding food you will want to eat, as well as attracting charitable insects that will feed on the annoying pests. Just be sure that the insects you bring in can coalesce with your honeybees.

Soil is also something you should pay attention to. You reap what you sow, and if you are sowing into a chemical fabric, the outcome will be anything but organic. By all means, use organic soil, and when you need a little kick to get things started, compost, mulch, and manure are Mother Nature's miracle grow.

Permaculture

By definition, permaculture is a system of development and cultivation intended to create a maintainable ecosystem that is self-sustaining and relies upon renewable resources. Permaculture and beekeeping go hand-in-hand because everything works together. The process of pollination by

FACTS ABOUT BEES

Bees can dance! Worker bees will wiggle their bodies in a figure eight motion to indicate the direction of a good food source to other bees.

- The area on a bee's leg where they carry pollen is called a pollen basket.
- Bees fly at an incredible speed for their small size. A honeybee's wings beat 200 times a second.
- Honeybees are the only kind of bee that will die if they use their stingers. This is because their stingers are non-retractable and when they inject it into something they also leave behind a piece of their abdomens.
- Bees have two different stomachs. One is for eating and one is for storing nectar that will be regurgitated into the combs to produce honey.
- A Queen bee can lay around 1,500 eggs a day.
- There are no other insects in the world that make a food like honey that humans can eat.
- Bees use their wings for more than flying. They frequently use their wings to fan their honeycombs (to dry nectar into honey) and the hive (to keep in cool).
- Bees are the only insect that produces food eaten by man.
- The scientific name for a honeybee is Apis Mellifara.
- Honeybees communicate within the hive by what is known as the waggle dance—yes, honeybees are dancers!
- It takes about 556 worker bees to produce 1 pound of honey, that's about 2 million flowers that get visited by the honeybee.
- The honeybee makes its distinctive buzzing sound by making 11,400 wing strokes per minute.

the bees creates thriving herbs, trees (such as apple), and flowers. Bees enjoy squash and bean blossoms, dandelion, and clovers, all of which are either beneficial for you as a gardener or are practically free. Be sure to raise flowers and plants that give nectar and pollen at different times of the growing and pollinating season, especially early and late periods. Perennials are helpful too because they return every season. Everything that you grow will also come to an end and die, which is where compost and mulch come in, giving the soil richness for next year.

Advice for Beginners

According to Mares, patience is the key. "Don't get discouraged," he urges, because bees "are wonderfully interesting." For the first year, both Mares and Berger recommend having a mentor. Knowing firsthand how important it is to have some experience before jumping in to the world and life of beekeeping, Berger says, "Someone who has bees and is willing to give you some pointers helps a lot. I did not have a mentor; I had to figure most things out for myself." "Spend time doing it," says Mares, because bees need and deserve a lot of attention.

Berger also suggests joining a beekeeping club. "When I started going to meetings the whole beekeeping world opened up," he explains. "Visit with your fellow beekeepers; you will learn a lot and gain valuable resources. Most of all, relax, enjoy the bees. Manipulating a hive is an art; knowing what you are seeing is a science."

Make sure you are involved with your bees. "Look for signs of overcrowding which will always lead to swarming," says Rowland. Some signs of swarming include: bees crowding on top of or outside of the hive, and honey or pollen in the brood nest which means the queen has no space to lay eggs. Population is important when it comes to beekeeping, and when swarming happens, "you end up losing half of your bees plus the queen," explains Rowland. Take measures like "start adding honey supers if you haven't already. Or if really crowded you may need to split the hive."

And when you get a colony, talk to your bees, recommends Robbie Minor, a knowledgeable beekeeper of two years who worked beside and learned from Les Crowder, a nationally renowned organic beekeeper of approximately forty years. Like a dog or cat, they know your voice, and most of the time will be calmer around you than if they were around a stranger.

Advice for Experienced Beekeepers

The best thing someone who has been beekeeping for a good number of years can do is keep up with the beekeeping lifestyle. That means, proposes Mares, keeping up with pests and viruses regarding your hives. You can "never know all there is to know about bees," says Mares, so keeping in touch with other beekeepers is always good too.

Keeping up with the beekeeping lifestyle also means you should "keep up on politics surrounding bees," recommends Berger. He also recommends that you "keep a subscription to *American Bee Journal* and/or *Bee Culture* magazine." However, if someone's goal is purely business-based, Berger's best advice is to "diversify your products," if possible.

Rowland suggests to "try something new in one or two hives each year. You may not make too many changes to your systems, but it will always add to your general knowledge base, and that is how we slowly improve the processes over the years."

FLOWERS BEES APPRECIATE

Although they smell effervescent and mesmerizing, flowers that have been genetically modified such as lilacs and tulips do not attract honeybees. They love natural wildflowers such as dandelions and clovers, as well as tree buds, squash blossoms, bean flowers, and fruit tree or bush blossoms. Favored seasonal flowers consist of alfalfa and goldenrod in the fall (which makes good provisions for the winter), and marigolds, bass wood, and clove in the summer.

The Bumblebee Conservation Trust suggests planting a bee-friendly garden instead of having a lawn. Planting a wildflower-rich meadow is ideal to provide habitat, especially if you plant nectar- and pollen-rich wildflower species. Children can get involved as well!

March–April: bluebell, borage, bugle, calendula, crocus, flowering currant, hyacinth, lungwort, pussy willow, rosemary, wild lilacs.
May–June: bee balm, cosmos, echionacea, chives, cotoneaster, everlasting pea, foxglove, honeysuckle, snapdragon, thyme.
July–August: buddleia, lavendar, marjoram, rock-rose, sunflower, heathers.
For Fall: plant late-bloomers like zinnias, purple asters, witch hazel, goldenrod.

Simple, single flower bloomers like daisies and marigolds are easy to plant and provide much-needed food for all kinds of bees. Designing a plan for replacing your lawn is a great way to save natural resources. like water, too.

JUDI HARVEST

*American artist Judi
Harvest (www.judiharvest.
com), whose work focuses
on the decimation of bee
populations, is keeping
beehives and making honey
near her glass-blowing
studio on Murano, Italy and
on the roof of her New York
studio.*

Photo courtesy of Judi Harvest

*Harvest creates
sculptures, paintings, videos, and installations inspired by the fragility
of life and the search for beauty. She believes there is no difference
between art and life.*

*Judi Harvest has had fifteen solo exhibitions in Venice. She has
an MFA in painting, having studied at Tyler School of Art in Rome,
Italy; the School of Visual Arts, New York and Urbino, Italy; New York
Studio School; and the Art Students League. From 1987–1992, she
lived and worked in Venice, Italy, where she studied glass making in
Murano. Concerned with the ecological issue of Colony Collapse, she
became a beekeeper in 2006. Her recent exhibition during the 2013
Venice Biennale, titled Denatured: Honeybees + Murano, united her
interests in the importance of bees to human survival and her love of
glass as a medium. Other solo exhibitions include Galerie Thomas, in
Munich and Caffè Florian, in Venice. For the 2015 Venice Biennale,
Harvest created the Room of Dreams installation with twenty-one
Murano blown-glass bed pillows placed on a round mattress base filled
with lavender from the Honey Garden.*

*Judi Harvest has work in the permanent collection of IVAM,
Institut Valencià d'Art Modern, in Valencia, Spain. Along with
exhibitions of her paintings and glass sculptures in Venice, she has
created three glass-based public artworks in the city: Fragmented
Peace (2003) and Luna Piena/Full Moon (2005), installed at the*

Vallaresso vaporetto stop, and Venetian Satellite (2006), first shown at the Caffè Florian in Piazza San Marco and currently on view in New York in the lobby of the West Chelsea Arts Building where she has her current studio.

About her ongoing exhibition in Venice, Harvest writes, "DENATURED: Honeybees + Murano is a site-specific exhibition in Venice conjoining two distinct, poetically similar, highly endangered systems of collaborative labor: honeybees and the hive (natural) + glass makers and the factories of Murano (social).

"Working with a beekeeper in 2013, I installed 4 hives in Murano in a garden I created for the honeybees. They are at the glass factory where I work on my art projects. At the same time, I created 100 unique glass Honey Vessel sculptures. I am currently working with the honeybees, harvesting honey, maintaining the Honey Garden and creating artworks inspired by the bees. The glass sculptures, paintings, and Breakfast with the Bees video were presented in Venice during the 2013 Biennale. DENATURED continues to bring environmental, social, and historical issues to an international audience."

How long have you been keeping bees?

I enrolled in a beekeeping class in 2006 after I heard about Colony Collapse Disorder and began keeping bees in New York City and Venice, Italy in 2013. I've spent nine years studying, researching and loving honeybees, and over two years taking care of my own. I currently have five hives in NYC on top of my studio building in West Chelsea overlooking the High Line and five in Murano, Venice, Italy. Every day I learn something new about honeybees and love and respect them even more.

What inspired you to begin as a beekeeper?

Honeybees are fragile and fundamental to our survival and food supply. They are the antennas of the environment. I want my art to help bring awareness to these tiny creatures and the bigger picture they represent. Honeybees need our help. I love flowers, honey, fruit, and gardens, and honeybees create beauty.

They are the most interesting and spiritual subject I have been inspired by as an artist. I have learned a lot from honeybees. Working with them keeps me in touch with nature and helps me maintain an inner balance in contrast with the hectic city and technology-filled life we are now living. Like the honeybees, we need to stop and smell the flowers and be nourished by them.

In the beginning, I hoped I could help bring awareness to the environment and the Murano glass factory. Now I realize how much the honeybees have helped and influenced me.

Any tips for beginners?

Have no fear, be calm when approaching a hive, take a beekeeping class, befriend passionate beekeepers. Relax, read, observe, be kind and gentle towards them, and they will reciprocate. Hang out with other beekeepers. Bee people are very special. Plant bee-friendly flowers and plants. Get involved from the ground up. Build a honey garden in an abandoned field; I did, and it changed my life. Give back to nature; there is a sense of urgency, and we cannot be passive. Bees need our help.

Wear white or solid colors and no perfume or scented products around the bees; otherwise, they may confuse you with a flower.

What type of hive do you use?

I use Langstroth hives in Venice and in New York. I use these as it is the type preferred by the beekeepers I have worked and studied with, Andrew and Norm Coté. They are fifth-generation beekeepers from Connecticut and New York.

The Langstroth hive was designed and patented in 1852 by Reverend Lorenzo Lorraine Langstroth and incorporates "bee space," which is ⅜ of an inch—the minimum space for a bee to move in. Honeybees build wax combs in the space of ⅜ of an inch or wider and ⅜ of an inch with propolis, which maximizes nutrients and minimizes space for hive invaders.

The Langstroth hive makes honey extraction easier as the frames are fully removable. Also the Langstroth hive allows for expansion in that additional boxes (supers) can be added. This prevents overcrowding and adds space for honey storage. I use a wooden frame with wax foundation in the Langstroth hives. I use all-natural materials in my hives and paint the exteriors with water-based, environmentally conscious paint. I try to make them attractive as honeybees recognize their hives by color and the scent of their queen. In Murano, I painted the hives the colors of the houses nearby where they forage on the island of Burano. There is a photo of this in my DENATURED 2013 catalog on my website.

I am an artist; the materials I use and expose myself to are always of highest quality and of the least possible danger to health—mine and others.

Certainly the honeybees have enough to worry about, pollinating over 350,000 crops. One out of three bites of food we eat requires honeybee pollination. My goal is to keep them safe and healthy.

Please relate a memorable experience you had as a beekeeper—humorous, instructional, tragic, or all of the above.
All of the experiences are memorable; every minute I spend with the honeybees is special. I go up to the roof every day to say hello to them and bring them fresh water. It is the highlight of my day.

The first time I saw the undertaker bees at work—removing the dead or dying honeybees from the hive and staying with them until the end—was most touching. I still get teary eyed when I observe this. I made a honeybee pool (small pebbles in a shallow bowl) for drinking water in Murano, and one bee fell in. Since they do not swim and I was watching, I thought to help her. At that moment, another bee landed, helped her out of the water, and fanned her with her wings until she dried off and flew away. Lifesaver bees and undertaker bees have really made an impact on me as it shows that honeybees have an instinct, to help and care for each other.

I have five hives on the island of Murano in Venice, Italy. The honeybees travel three miles in each direction every day foraging, often over large bodies of water. When they are tired, they hop on the vaporetto (water bus) and get off where they live in Sacca

Serenella in the Honey Garden. I have filmed this a few times, amazed to observe their innate sense of direction. This is on my website in the video *Sogni* from my 2015 Room of Dreams exhibition.

What products do you make?
With the help of the honeybees, I make Murano and High Line Honey and artworks inspired by the bees. It is important to buy honey directly from a local beekeeper so you know it is real and pure honey. It costs more, and it is worth it. There is a lot of fake honey being sold in stores, made in China and extremely unhealthy.

I may try creating candle and beauty products from the beeswax and honey in the future. I am currently working on a line of "Queen Bee Murano" glass jewelry.

How many hives do you recommend someone begin with?
Four if you have help; two if not.

Where do you get your bees?
From Andrew and Norm Coté in New York and from Elio Mavaracchio in Sant Erasmus in Venice.

Where is a good spot to put a beehive?
Anywhere with sunlight, flowers, and space. Each hive needs 5 feet of space around it. They do very well on rooftops in New York City and Paris. Although it may be more beautiful to place your hives in a garden setting, they can be placed securely on a rooftop as the honeybee is up and out at sunrise and flies upwards towards the sun and in search of flowers.

They will hang around a garden and their hives, but they like foraging and reporting back to their sisters where the food source is. They do this by dancing. Honeybees are the only insect that communicates by dancing. I went to a dancing school to understand this concept and now also consider dance one of the highest forms of communication. In 1946, the Austrian scientist, Karl von Frisch observed honeybees dancing while studying the effects of color and honeybees. In 1973, he won the Nobel Prize for his discovery.

The honeybees have two specific dances: the Waggle Dance when the food source is far from the hive and the Round Dance when the source is closer. The dances give accurate directions as to where and what kind of flowers are nearby. The Waggle Dance forms the figure eight shape. The pollen collects on their furry bodies and the scent of the flowers is shaken as they dance to inform the other honeybees where the flowers are. By dancing, the honeybee communicates distance, direction, and the quality of the food source. Quality is communicated by the food source stimulation, resulting in a more enthusiastic dance. The honeybee's dance floors are at the entrance of the hive, so they do not waste time informing the others where the flowers are.

How do you attract bees?

Plant a bee-friendly garden: Anything purple or blue—lavender, sunflowers, sage, wisteria, roses, clover . . . . *No neonicotinoid-based pesticides, ever*. Be very careful when purchasing plants for your garden that they are not treated with pesticides already. Pesticides are very unhealthy for the honeybees and us as we eat the food they pollinate.

What do you do when bees or the whole hive become sick?

I have been most fortunate; my hives have survived extreme New York winters and Venice humidity. Either I or a fellow beekeeper opens the hives and checks on them once every week or two in the spring, summer and fall. If I have questions, I call my beekeeper friends and ask advice.

There are natural treatments for all of the possible bee illnesses. For mites, my beekeeper uses powdered sugar sprinkled over the hive frames; the honeybees fly out and shake it off, along with the mites. It looks like a snow flurry. I am a member of New York City Beekeepers, and we share information at monthly meetings and by email. Like a real hive, one honeybee cannot work alone. There are beekeeper groups now all over the world, a very interesting and helpful resource.

What flowers do you find bees are most attracted to?
Anything purple, especially lavender, sage, sunflowers, and roses. They love fruit trees, too. In Murano, I planted nectarine, cherry, pear, and pomegranate trees. They all had fruit in the first year, thanks to the honeybees. On my terrace in New York, the honeybees love visiting the white wisteria tree. This can be seen in my "Breakfast with the Bees" video on my website.

It takes 5 million flowers to make a pint of honey. We need to plant more flowers and take time to smell them. The honeybees remind me of this, and I am most grateful.

Keeping Bees Happy and Healthy through All Seasons

Bees, although it may not seem like it, are very susceptible to disease and illness. Keeping your bees on an organic lifestyle aids their well-being and benefits their health. Because they are not covered in or consuming chemicals, sickness will be a smaller part of both your life and their lives. Checking on your bees is something that should be done frequently. You'll catch problems sooner and be able to determine their progress if they are recovering. If there is an illness or mite infestation, be sure to take the necessary measures, and possibly change out the queen or use other techniques to rid the bees of the problem.

In order to keep your bees happy, think of things that make their lives easier. If there are flowers in your garden that they strongly favor, they won't have to travel in order to forage for pollen. Also, in preparation for the harsh winter months, make sure the hives are secure and can function properly until spring blooms and they can be freed from the hive once again. Location is another thing that impacts the bees' happiness. Bees don't like to be disturbed frequently by things such as children, animals, or rough weather, so make sure the hive is somewhere that keeps the bees away from human and animal activity in addition to having enough shelter from Mother Nature.

WHY SMOKE?

The use of smoke in correspondence to beekeeping has been around since the Egyptians. Ancient murals on the walls of Niuserre's sun temple, part of the Old Kingdom, have been found depicting beekeepers blowing smoke into the hives in order to extract the honey combs peacefully. They had good reason to do this. When encountering smoke, bees instantly believe there is a fire, in which case they dine on honey in a frenzy of anxiety and haste. Because they are full, they become lazy, calm, and less likely to sting. Some people believe it is cruel not to use smoke due to the fact that once a bee stings, it dies. Although smoke is not the best, it practically saves those bees' lives who otherwise would have died as a result of their temper. However, some beekeepers have managed to work without smoke and still keep a low count of fatalities every time they have to disrupt the hive, but if you are a beginner or feel like you do not want to take the risk, a little smoke will not hurt the bees.

HUMANS, BEES, AND SOCIAL BEHAVIOR

Our society seems original in that people are divided into categories of function, and when we have to, we communicate with many groups we would not normally reach out to. By doing this, we create circles of interactions that a majority of other species lack. Those who are in need are helped—to some extent—and we care for the ill, elderly, and unable. There is a moral system we, as a group, try to follow, and although we do not always follow said moral system, we are capable of doing so. Bees, no matter their size and intelligence, mirror our social behavior more accurately than would be assumed. These busy creatures clean other bees to rid them of harmful things such as parasites and take turns in feeding each other. They both establish and strictly maintain a monarchy while taking care of all infants and keeping their colony safe by fighting off enemies. Also like humans,

bees will sometimes strike down a weaker colony in order to better their own in a time of dire need. When two colonies are moderately weak, they are often combined to create a stronger, more stable single colony. Rings of loyalty are entrenched, and the two bee colonies soon begin to work together. The resemblance is sometimes fuzzy in that we care what we wear when all the bees have are their stripes, but the social behavior between the species should not go unnoticed.

PART II
FRUITS OF
THE BEES' LABOR

How and What to Harvest from Your Hive

"Handle a book as a bee does a flower, extract its sweetness but do not damage it."

—John Muir

Harvesting Honey

There is nothing sweeter than reaping what you have sown or, in terms of bees, harvesting what you have helped to create. But before getting right down to slurping up the sweet gold, you have to pay heed to one thing of great importance. You must make sure not to contaminate the capped and already cleansed honey with honey from uncapped cells. This uncapped honey, or green honey, contains a large amount of yeast due to a high level of water saturation, making it unsafe to eat. Also, the texture is often times gooey, and it does not taste good.

When it comes to honey, the rule quality over quantity comes into play. The color of the honey reflects where the bees are located, as well as the season in which the honey was made. Obviously, different flowers bloom at different times, so the flavor and color depend on what flowers were pollinated during that time. The National Honey Board has set six different official honey colors to categorize honey types, even though there are over three hundred types. The six colors are water-white, white, extra-light amber, light amber, amber, and dark amber. Usually, people just guess by visual observation, but there are tools that digitally calculate how much sunlight the honey absorbs.

One way to harvest honey is by harvesting it one comb, or top bar or frame, at a time. This is a homemade way of extracting honey and can be done in less than thirty minutes, depending on the size of the comb from which you are harvesting the honey. This way allows for honey created from the same pollen source to stay together, creating a vibrant and solid jar of honey. Of course, you have to know which combs have been filled with honey made from the same flowers, so keep an eye on your bees and make a mental or physical note of where they stored which honey. Collecting honey one comb at a time is also great for beginners because it gives them a chance to harvest some honey without the stress of having to do the whole hive at once. The bees are not disturbed very much either, which means you won't have to gear up all the way and remove the bees from the hive. Some may say it builds trust between the bees and the beekeeper because the two parts of the relationship are happy due to the fact that honey is being shared. However, this process only produces a little bit of honey at a time, and in order to harvest all of the honey, you will need to repeat this process many times, although it can be and usually is spread out over a few days or even weeks.

There are three steps to harvesting honey from a single comb at a time, and all you need are two one-quart mason jars with the jar rings (but not the lids), a piece of mesh with very fine screening that is big enough to stay secure at the top of one mason jar, and duct tape.

1. Cut the comb off of the top bar or frame, depending on what type of hives you have. Do this using an uncapping knife, and run it down both sides of the comb. Next, mash together the honey and wax in a bowl until the wax is well broken up and the substance has the same texture all the way through.
2. Pour the wax and honey into a one-quart mason jar, and without the lid, screw the jar ring on. Using another one-quart mason jar, place a piece of fine-screened mesh on top, and secure it by screwing the jar ring on. Mosquito netting works well but any sort of cloth does not. Spray the mesh with a little bit of water to decrease the surface tension that will be brought on by the thick flow of honey.
3. Now, it is time to make a liquid gold hourglass. Turn the empty jar with mesh upside down, and fit it atop the second mason jar filled with honey. Use duct tape to secure the two jars, and make sure it is tight enough so no honey can trickle through. Then, flip the hourglass over, and let the honey trickle through the mesh. Place it in a warm location for approximately an hour. When you return, the clean honey will be separated from the wax, and it will be ready for consumption.

However, if you are someone who likes to get things done all at once, you can harvest all of your honey in five easy steps. You will need gear that will protect you from any angry bees that weren't subdued by the smoke. Preferably, you'll have a hat and veil, durable gloves, a jacket and/or a bodysuit, a brush with a wooden handle or an electrical blower (to gently remove lingering bees from the frames), something to uncap the cells such as an uncapping knife, a manual or electrical honey extractor, and jars or other containers to put your honey in.

1. Approaching the beehive from behind, puff smoke around the entrances. Take the lid off the hive, and puff more smoke on the opening you have created. This causes the bees to go deeper into the hive. Next, strip away the inner cover. This may be difficult to do because the bees may have sealed the cover shut with propolis, a resin they make after pollinating trees. It is good for the bees, says Minor, because they use it as sort of a sticky velcro that pulls away all the bad stuff on them when they enter the hives. (It is also beneficial to humans, but no need to worry about that right now.)
2. It is important to remove the bees from your hive no matter how much honey or how many pieces of the hive you intend to remove. There are many different ways you can go about this. If you want to travel along the gentle and organic route, use a brush with a wooden handle to brush away the bees from each frame. There are the other mechanical-based options involving gas or electrically powered blowers. Whichever method you choose, once the bees are removed from each frame, set it aside (away from the bees) until it is ready to go through the extractor.
3. To uncap the honey combs, take an uncapping knife, scratcher, or fork, and run it down both sides of each frame.
4. Honey extractors come in manual and electrical versions, so you can choose which you prefer as well as which one fits into your budget. Just place the frames into the extractor, as many as will fit properly, and coax the honey out of each cell by spinning the frames. The honey then drips down the drum wall, collecting at the bottom and flowing through a spigot that should be open for the honey to travel through and exit. Be sure to strain your honey multiple times using cheesecloth to make sure there aren't any unwanted textures in the smooth sweetness.

5. Now comes the easy part: bottling your honey! Be sure to sterilize the bottles you intend on using, so nothing contaminates what you have worked so hard for.

Beeswax Candles

HISTORY OF BEESWAX CANDLES

Candles have been made with beeswax in early American culture and in European culture since the Middle Ages, when the product was introduced to Europe as an alternative to burning tallow (animal fat). Previous to this, the Egyptians made wicked candles in 3000 BCE, by dipping rolled papyrus in melted beeswax. When these natural candles were brought to Europe, they were a widely accepted alternative to tallow because they emitted a cleaner scent, without producing a smoky flame. They were also widely used for church ceremonies, which included the Paschal candle, or "Easter candle," as well as being used to light the Roman Catholic and Orthodox Churches throughout European history. When beekeepers extracted honey, they would use a cheesecloth to press the honey out of the beeswax honeycomb. With all the extra wax, candles would be formed to later burn during the uneventful winter months, giving off the same soothing and comforting aroma candle-burners enjoy in their homes today.

BEESWAX CANDLES VS. UNNATURAL TAPER CANDLES

Bee products are a fantastic alternative to unnatural and often harmful products used in many homes. For instance, many households burn the classic taper candles, which are widely available in most home decor sections of retail businesses; however, the taper candle isn't always the healthiest choice. They are made out of paraffin wax, which is bad for the environment, because it comes from petroleum, which is not a renewable resource. Additionally, when burning this wax, it releases toxins in the smoke and soot, making the air in your home unhealthy. An alternative to this is the beeswax candle, which contains no additives, is naturally sweet-smelling, and a cleaner choice for the environment. When burning beeswax candles, negative ions are produced which attracts pollutants such as dust, odors, viruses, and bacteria; as a result, beeswax candles actually clean the air in your home, instead of releasing harsh chemicals. Not only are beeswax candles a healthier alternative for the air you're breathing within your home, but with their unique and natural texture, they are a beautiful interior decorating choice as well as a great way to support your local beekeeper.

MEDICINAL USES

During ancient times, honey was used to cure many ailments, but not until recently have doctors and other health seekers revisited using honey as a well-rounded medicine. Due to many people beginning to show great resistance to the common antibiotics used to cure some diseases and

illnesses, doctors needed something else, and they rediscovered honey. As a result, apitherapy was recently developed and enables patients to be treated using bee-made products.

Organic, unheated honey has an array of antibiotic properties— stemming from the naturally occurring hydrogen peroxide in honey— that can fight more than one hundred types of pathogenic bacteria. It is hygroscopic, meaning it takes moisture from an area and dehydrates the bacteria. This stunts the growth of the bacteria, and because of decreased pH levels in the honey and an increased amount of sugar, microbes cannot grow further.

So, honey, with all its antibiotic characteristics, can heal different wounds and conditions. It accelerates wound healing by creating a good environment for tissue growth and a bad environment for microbial growth. It works well when trying to heal stomach ulcers, bed sores, burns, and other infections of the skin. This is also partially because it acts like a barrier, so as it heals the wound, nothing from the outside can perturb the process. Honey reduces inflammation and can be used on skin grafts and where the skin was taken. If ingested, it may help repair intestinal mucosa that was ruined or damaged by peptic ulcers or gastritis. Furthermore, in all case studies, it has never been reported that microbes resisted the antibiotic properties of honey.

Although different types of honey have different antibiotic properties, manuka honey is a powerhouse due to its diverse healing skills. Made from the pollen of mānuka bushes that are native to New Zealand, a variety of medical problems can be treated using this one honey. Like all honey, it accelerates healing in wounds, boosts energy, and makes skin healthier, but it can also help heal venomous ulcers, clear up acne and eczema (face masks made with the honey are applied to the site), and provide relief when you are sick. The honey empowers the cells that create immunity to some viruses, as well as annihilates strep bacteria, so your sore throat goes away faster. It can also help cancer patients who are going through chemotherapy because it acts as an anti-inflammatory agent and cools their throats.

By chewing or sucking on a manuka honey product, you can reduce plaque buildup, resulting in decreased gingivitis symptoms or tooth decay. Small intestinal bacterial overgrowth (SIBO) can be treated using this certain kind of honey, as well as low stomach acid and acid reflux by its ability to help equate the digestive system. Staph infections can also be kept at a low level by using manuka honey. Due to the increased resistance to many of the usual treatments used, staph infections have been spreading rapidly, leading researchers to seek a more natural option. The antibiotic properties of the honey take away the majority of the power in the more solid genes of staph bacteria, so after a while, the bacteria becomes weak and the illness decreases in intensity. Patients then do not have to go to extreme measures like surgery to insert fake joints and tubes to live. Those with staph infections just have to apply the honey to any cuts or infectious openings to keep a staph infection down.

Those who have allergies, whether they be intense or minor, can benefit by consuming manuka honey. Symptoms will be reduced, and some may even disappear, which makes it so you don't have to take as much medicine. This type of honey also acts as a natural sleep agent when added to warm milk before bed. It causes the body to easily release melatonin, the chemical released by the brain to induce sleep.

The best way to see any kind of health benefits, besides the cases that call for the honey to be applied to a certain area, is to ingest one to two tablespoons of manuka honey a day. If you don't like the taste or if it seems too sweet for your liking, you can add it to yogurt, your favorite tea, or even spread it on toast.

RECIPES

"The fruit of bees is desired by all, and is equally sweet to kings and beggars and it is not only pleasing but profitable and healthful; it sweetens their mouths, cures their wounds, and conveys remedies to inward ulcers."

—Saint Ambrose

In this section, at the end of our journey with the flight of the bees (so to speak), we will explore different ways you can use honey and things you can make with honey.

My lifelong love of honey has yielded a respect for and fascination with bees. A daily spoonful of honey in my tea is just one thing I do with nature's sweetener. There are many uses for the honey that you either collect from your backyard hives or purchase from local growers.

BREAKFAST

Breakfast Shake

Ingredients

1 tbsp honey
1 ripe banana
¼ cup orange juice
1 cup fresh or unsweetened frozen strawberries

Directions

Combine all ingredients in a blender. Blend until well mixed. More orange juice may be added for a thinner shake.

Makes 1 serving. No fat, no cholesterol.

Honey Quinoa Breakfast Bake

Ingredients

1 cup quinoa (uncooked)
1 tbsp cinnamon
2 cups mixed frozen berries
½ cup coarsely chopped nuts
2 eggs
¼ cup honey
2 cups milk

Directions

Preheat oven to 350°F. Spray an 8x8-inch baking dish with non-stick spray. In a small bowl, stir together uncooked quinoa with cinnamon, making sure it is coated completely. Pour quinoa over bottom of prepared dish. Scatter the berries and nuts on top of quinoa, making sure to spread evenly. In a small bowl, lightly beat the eggs. Add the milk and honey and whisk together. Pour egg mixture on top of quinoa and fruit.

Bake 1 hour or until the breakfast bake only has a small amount of liquid remaining. Serve warm.

Honey-licious "Energizers"

Ingredients

1 cup fresh dairy butter
1 cup honey
1 tbsp freshly grated orange zest
3½ cups quick cooking organic oats or large flake
⅛ tsp sea salt

Directions

Melt butter, BeeMaid honey, and orange zest in large pan over medium heat, to create syrup. Simmer for 2 or 3 minutes to release the orange flavor from the zest. Remove from heat and stir in oats; and salt mix well. Line 9x12 glass baking dish with parchment paper, tip oat mixture into it, and fork over lightly to spread.

Bake at 325°F for 20 to 25 minutes on middle rack until deep golden brown.

Let cool. Lift parchment to board and slice. Good keeper, but it never gets a chance to sit around because it is so yummy! Excellent for kid's lunches, hiking, and picnic baskets.

VARIATIONS:

Can also add dried cranberries, raisins, chopped nuts, and sunflower or pumpkin seeds. Use your imagination and make this recipe your own.

After trying each variation, my family votes for the original version every time. We think that is because of the full flavor that comes from using BeeMaid honey. When cooked, it has a caramel, almost butterscotch flavor. The orange zest brings out the floral notes in the honey and all in all, it is hard to beat such an easy recipe. When each ingredient is of the highest quality, then the end result is simply outstanding.

Comfort Food

Ingredients

Bread of your choice toasted slightly (raisin is specially good)
2 tbsp honey
1 cup warm milk

Directions

Pour the honey and milk over toast for a bedtime snack that is healthy or just plain comfort food for any age.

Grab & Go Breakfast Bars

Ingredients

⅔ cup honey
¾ cup almond butter
4 cups whole grain cereal or granola

Directions

1. In 4-cup microwave-safe container, microwave honey at HIGH (100%) for 2 to 3 minutes or until honey boils.

2. Stir in almond butter; mix until thoroughly blended.

3. Place cereal or granola in large bowl. Pour honey mixture over and combine thoroughly.

4. Press firmly into 13x9x2 baking pan. Let stand until firm. Cut into bars.

APPETIZERS & SALADS

Fruit Salad with Honey Lime Dressing

Ingredients

½ cup honey
½ cup lime Juice
Pinch nutmeg or cinnamon
4 cups sliced fruit (a combination of berries, apples, melon, etc.)

Directions

In blender or food processor, combine honey, juice, and seasoning; blend until smooth. In medium bowl, toss fruit with dressing and chill until ready to serve.

Almost Honey Glazed Popcorn

Ingredients

1 cup butter
¾ cup honey
¾ cup brown sugar
½ tsp salt
1 tsp baking soda
1 tbsp vanilla extract
4 quarts popped corn
1 cup toasted silvered almonds

Directions

Preheat oven to 250 degrees. In a medium saucepan, melt butter then add honey, brown sugar, and salt.

Bring to a boil, stirring constantly, then reduce heat and continue to cook for 3–5 minutes, stirring occasionally. Remove from heat and add baking soda and vanilla extract, stirring quickly.

Pour mixture over popcorn in a large bowl, mixing to coat. Add almonds; mix thoroughly.

Pour onto 2 greased baking sheets and bake for 45 minutes, stirring every 10–15 minutes. Cool on sheets, break into pieces, store in an airtight container.

Grilled Veggie Honey Flatbread

Ingredients

¼ cup honey
3 tbsp white wine vinegar
1 tbsp fresh thyme or basil, finely chopped
¾ lb. small zucchini or yellow summer squash, cut lengthwise in half
2 large red, yellow, orange, or green bell pepper, halved and seeded
2 tbsp + ½ cup toasted wheat germ
1½ cups reduced-fat baking mix
⅔ cup fat-free or low-fat milk
1 cup tomatoes, seeded and chopped
1 package (4 oz.) feta cheese, crumbled
Nonstick cooking spray

Directions

Makes 4 Servings
In a small bowl, combine honey, vinegar, and thyme; mix well. Place zucchini slices and bell peppers on oiled grill. Grill over medium-hot coals 20–25 minutes, turning and brushing with honey mixture every 7–8 minutes. Remove from grill; cool to room temperature. Coarsely chop.

Heat oven to 425°F. Lightly spray a cookie sheet or jelly roll pan with cooking spray; sprinkle with 2 tablespoons wheat germ. In a large bowl, combine baking mix, ½ cup wheat germ, and milk; stir with fork until thoroughly combined (mixture will be moist). Turn dough out onto a floured surface. Knead, adding additional flour as needed, until dough is no longer sticky. Pat into a 12x8-inch rectangle on a cookie sheet. Top with chopped grilled vegetables, tomatoes, and cheese. Bake 18–20 minutes or until crust is golden brown. Cut into squares and serve immediately.

Whipped Honey Goat Cheese on Grilled Crostini

Ingredients

Multi-grain baguette, cut on the bias into ¼" slices
Olive oil
2 garlic cloves, cut in half
4 oz soft mild goat cheese, at room temperature
2 tbsp whipping cream
3 tbsp honey
Fresh cracked pepper
Fresh chives, rosemary, or other herbs

Directions

Preheat a barbecue over medium heat. Brush one side of each baguette slice with olive oil. Rub side with olive oil with garlic. Place baguette slices, olive oil side down, on barbecue and grill until lightly charred. Remove from heat and let cool.

Combine goat cheese and cream in the bowl of an electric mixer outfitted with the whisk attachment; process until well blended. Add honey and process until combined.

To Assemble:

Place mixture into a piping bag fitted with a large star tip. Squeeze goat cheese mixture onto baguette slices. Add a grinding of fresh black pepper. Top each with fresh herbs.

Cover and chill until ready to serve.

Honey Cucumber Salad

Ingredients

3 medium cucumbers, thinly sliced and halved
¼ cup honey
½ cup white balsamic vinegar (can also use white wine vinegar)
¼ cup water
2 tbsp red onion, diced
Salt

Directions

Place cucumbers in bowl and sprinkle with salt. Toss and set aside. In a small mixing bowl, stir together honey, white balsamic vinegar, water, and diced red onions. Pour the mixture over the cucumbers and toss. Allow the salad to marinate in the refrigerator about 1 hour prior to serving.

MAIN DISHES

Honey Garlic Meatballs

Ingredients

2 lbs ground beef
1¼ cup fine breadcrumbs
¾ cup milk
½ cup onion, finely chopped
2 eggs
¼ tsp pepper
2 tsp salt

1 tbsp butter
4–5 cloves garlic, crushed
¾ cup ketchup (or ½ ketchup and ½ bbq sauce)
½ cup honey
¼ cup soy sauce or tamari

Directions

Meatball:

Combine meat, breadcrumbs, milk, onion, eggs, salt, and pepper in bowl. Mix well with fork. Form into 1-inch balls.

Place balls in single layer on cookie sheet. Bake 500°F for 10–12 minutes. Drain well.

Sauce:

Melt butter in saucepan and sauté garlic until tender. Add ketchup (ketchup bbq sauce mixture), honey, and soy sauce. Bring to a boil.

Place meatballs in greased casserole dish, pour sauce over top.

Bake 350° for 20 minutes to ½ hour.

Honey Grilled Coconut Prawns

Ingredients

¾ cup Ortega white wine or similar
2 (6–8 tbsp) fresh limes, juiced
6 tbsp honey
2 tbsp extra virgin olive oil
2 tbsp freshly chopped cilantro
½ tsp sea salt
White pepper to taste
2 cloves garlic, minced
1¼ lbs large and peeled and deveined shrimp
¾ cup unsweetened shredded coconut
Mixed baby greens on platter
½ cup coarsely chopped toasted cashews

Directions

Whisk together wine, lime juice, honey, olive oil, cilantro, salt, pepper, and garlic in a bowl.

Set aside ¼ cup marinade for later and add shrimp to remaining bowl of marinade.

Cover and refrigerate for 15 minutes.

Remove shrimp from marinade and thread shrimp onto skewers. Discard marinade.

Gently press each side of skewer into dish of coconut.

Grill over medium heat, turning until completely pink and coconut is toasted.

Place skewers on greens dressed with reserved marinade and sprinkle with warm toasted nuts.

Makes 4 entrees.

Great for a light meal or summer patio luncheon. Ortega white wine can be substituted with Vinho Verde or Chardonnay.

Honey Glazed Chicken Wings

Ingredients

3 lbs chicken wings
⅔ cup light soy sauce
½ cup honey
2 tbsp canola oil
2 tsp five-spice powder
3 cloves finely chopped garlic
Non-stick cooking spray

Directions

Cut the chicken wings at the joints, and discard tips.

Place the wings in a large plastic zipped lock bag. Pour all of the remaining ingredients into the bag. Shake up the bag to mix well, and refrigerate for 2 hours. Shake occasionally. Preheat oven to 375°F.

Place the marinated chicken wings onto a large baking tray, that has been sprayed with non-stick spray.

Place tray in the preheated oven, and bake for 1 hour, turning them over halfway through cooking time.

Serve with a green or Caesar salad, and a nice glass of your favorite wine.

Honey Mustard Tarragon Chicken

Ingredients

6 boneless chicken breasts
¾ cup honey
¾ cup prepared mustard
2 tsp tarragon

Directions

Spray a 9x13 baking dish with cooking oil.

Put chicken into pan. Stir together remaining ingredients and pour over chicken, cover and bake at 350°F for 1 hour.

Serve with wild rice and veggies.

Sweet & Sour Chicken

Ingredients

2 medium onions
4 carrots
1 whole green and red peppers
6 chicken legs (thighs attached or separate)
¼ cup all purpose flour
½ tsp salt and pepper
2 tbsp vegetable oil
½ cup orange juice
¼ cup honey
¼ cup soy sauce
¼ cup tomato paste
1 tbsp cornstarch
3 cloves minced garlic

Directions

Cut each onion into 6 wedges, cut carrots into ½ inch chunks, and seed, core, and cut peppers into 1 inch cubes. Scatter vegetables in large roasting pan (with lid), cover and roast at 400°F until slightly tender, 10–15 minutes.

Meanwhile, skin chicken.

In large shallow dish combine flour, salt and pepper. Dredge chicken in this mixture. In large non-stick skillet, heat 1 tbsp vegetable oil over med-high heat, brown chicken in batches adding remaining oil as necessary—about 8 minutes. Arrange chicken on top of vegetables.

In large measuring cup whisk together orange juice, honey, soy sauce, tomato paste, cornstarch, and garlic. Pour over chicken and vegetables. Cover and roast again for 20 minutes. Remove lid and roast, basting occasionally until chicken is golden and juices run clear (about another 20 minutes).

Makes 6 servings. Serve with steamed rice.

DESSERTS

Bittersweet Chocolate Frosting

Ingredients

¼ cup clover honey
8 oz. 60% cocoa bittersweet chocolate, coarsely chopped
1 cup heavy whipping cream
2 tbsp seedless raspberry jam, optional

Directions

Combine honey and chocolate in a medium bowl; set aside. In small, heavy pan, heat whipping cream over medium heat until bubbles just begin to form. Pour over honey-chocolate mixture and allow to stand for 2 minutes. Stir until smooth; cool. Refrigerate until chilled, 1–2 hours.

With an electric mixer, beat chocolate mixture until frosting is fluffy.

Crustless Pumpkin Pie

Ingredients

3 eggs
½ cup honey
½ tsp ginger
½ tsp nutmeg
½ tsp cinnamon
½ tsp salt
1½ cups cooked pumpkin
1 cup evaporated milk, undiluted
½ cup whipping cream
1 tbsp berry sugar
¼ tsp vanilla

Directions

If using uncooked pumpkin, cook down to 50% volume and puree.

Preheat the oven to 325°F. Beat the eggs slightly. Add honey, ginger, nutmeg, cinnamon, salt, and pumpkin. Mix well, then add evaporated milk and mix. Bake for 50–60 minutes until an inserted knife comes out clean. Cool.

Add berry sugar and vanilla to the whipping cream. Whip the cream and serve pie slices with the whipping cream.

Note: a piecrust is optional.

Honey Almond Butter Squares

Ingredients

⅔ cup honey
¾ cup almond butter
4 cups granola cereal mix

Directions

In 4-cup microwave-safe container, microwave honey at high (100%) 2–3 minutes or until honey boils.

Stir in almond butter; mix until thoroughly blended.

Place granola in large bowl. Pour honey mixture over granola and combine thoroughly.

Press firmly into 13x9x2 baking pan. Let stand until firm. Cut into squares. Makes 36 squares.

Honey Carrot Snacking Cake

Ingredients

½ cup butter or margarine, softened
1 cup honey
2 eggs
2 cups finely grated carrots
½ cup golden raisins
⅓ cup chopped nuts (optional)
⅓ cup orange juice
2 tsp vanilla
1 cup all-purpose flour
1 cup whole wheat flour
2 tsp baking powder
1½ tsp ground cinnamon
1 tsp baking soda
½ tsp salt
½ tsp ground ginger
¼ tsp ground nutmeg

Directions

Cream butter in large bowl. Gradually beat in honey until light and fluffy. Add eggs, one at a time, beating well after each addition. Combine carrots, raisins, nuts, orange juice, and vanilla in a medium bowl. Combine dry ingredients in separate large bowl. Add dry ingredients to creamed mixture alternating with carrot mixture, beginning and ending with dry ingredients. Pour batter into greased 13x9x2 inch pan. Bake at 350°F for 35–45 minutes or until wooden pick inserted near center comes out clean. Makes 12 servings.

Recipes for Health and Beauty

Green Honey Glow Mask

Ingredients

4 cups fresh spinach
1 piece (1-inch) ginger
1 cup fresh mint
3 tbsp honey
1 ripe banana
2 egg whites

Directions

Rinse spinach thoroughly in colander. Cut and peel ginger, set aside. In food processor or blender combine spinach, mint and ginger. Blend on low setting. Add honey and banana and blend until liquid consistency. Add egg whites and blend until all ingredients are mixed thoroughly. Transfer to porcelain bowl or glass dish. On clean skin, apply a small amount of Green Honey Glow to entire face and neck. Apply using a fan brush or finger tips. Allow to remain on skin for 15–20 minutes. Rinse and apply appropriate moisturizer. Store covered in refrigerator for up to one week.

Honey Hair Conditioner

Ingredients

½ cup honey
¼ cup olive oil (use 2 tbsp for normal hair)

Directions

Mix honey and olive oil. Work a small amount at a time through hair until coated. Cover hair with shower cap; leave on 30 minutes. Remove shower cap; shampoo well and rinse. Dry as normal.

Honey Body Scrub

Ingredients

½ cup lemon juice
1 cup honey
2 tbsp coconut oil
2 tbsp olive oil
2 cups white sugar

Directions

Combine lemon juice, honey, coconut oil, and olive oil in a large bowl. Slowly add sugar and stir to combine all of the ingredients. Mix, and add more sugar as needed.

Massage the scrub all over body and your face. Leave it on for a few minutes.

Rinse off with damp cloth or in the shower. Towel dry. Adjust oils and sugar as needed.

Honey Hot Ginger Tea

Ingredients

2–4 inches raw ginger
4–6 cups water
½ lime, juiced
4 tbsp or more honey

Directions

Peel the ginger and slice thinly to maximize the surface area. This will help you make a very flavorful ginger tea. Boil the ginger in water for at least 10 minutes. Remove from heat and add lime juice and honey to taste.

**For a stronger more flavorful tea, allow boiling for 20 minutes or more, using additional slices of ginger.

Beekeeping Success Stories

"Life is the flower for which love is the honey."

—Victor Hugo

Among the Angels of Agriculture

Local beekeeper, Mike Palmer, expert in his field
by Elodie Reed (used with permission from the *St. Albans Messenger*)

Mike Palmer, 65, of St. Albans, Vermont, works on catching and caging queen bees in a field.

Among the buzzing, crawling, and flying bees in a hot, sunny field behind a farm on Lower Newton Road earlier this week, Mike Palmer spoke about his passion.

The 65-year-old St. Albans resident, owner of French Hill Apiaries, has been working with honeybees for some forty years now, and is considered an expert in the field.

Palmer keeps seven hundred production hives and many more mating hives, producing thirty tons of honey per year. He also mates and sells his own queens, tours internationally to give presentations on beekeeping, and has no shortage of YouTube videos he's made on the subject.

"It's all I do. They've captivated me," said Palmer of bees and his work with them.

AN INDUSTRIOUS WORKER

To start, Palmer was a novice like most other beekeepers when he first began in 1974. He is originally from Long Island, N.Y., but stayed in Vermont after attending an agriculture program at the University of Vermont. Upon graduating, Palmer quickly found that a sugarbush was too expensive for him to buy, so he decided to work with bees. His beekeeping skills are self-taught.

"You can do anything for forty years and learn it," Palmer said.

Palmer builds most of his own boxes and frames on which the bees produce his honey. His apiaries, or bee-yards, are spread around the area, divided between honey production and mating hives.

As part of a philosophy he calls "the sustainable apiary," Palmer breeds his own queens each spring instead of buying them from southern state producers such as those in Georgia.

"They just don't seem to work here," he said. Palmer then added, "We should be able to grow our own bees. Raising bees is no different from raising corn, raising sheep."

The breeding process begins in late spring in "nucs," or smaller "nucleus" hives, where queens are grown and caught in the summer, and are caged and sold or used in Palmer's production hives for the remaining warmer months.

The queens die in the winter, though not before leaving behind queen cells that stay viable in winter and grow once spring returns. Palmer overwinters his nucs in order to allow this to happen.

Palmer shares his success with others, giving instructional presentations on sustainable apiaries once a month or so. He travels around Vermont, to other states, and even to other countries, such as England, for these talks. He was in the United Kingdom last fall, and will be traveling there again during the winter.

Palmer is also involved with the Vermont Beekeepers Association and the Franklin County Beekeepers Club.

A SWEET CONNECTION

It's clear that, in addition to enjoying the honey and income his bees bring him, Palmer really likes the bees themselves.

"They're fascinating," he said. When selecting queens or other bees, he chooses those that are good tempered, since they are the type of bees he'll then get down the line.

"It's all genetic," said Palmer.

Just because he picks out friendly bees doesn't mean they don't sting when irritated. As to be expected from a man who picks up the creatures by their wings all the time and is constantly in their presence, Palmer's no stranger to bee stings. "I've gotten stung I don't know how many times," he said as he showed his fingers, which are lined with tiny black sting spots.

Palmer made the point, though, that people shouldn't fear honeybees. "There's nothing to be afraid of," he said. "They do all that work for us," he added, referring to both honey making and pollination for much of our agriculture.

"Some people call them the angels of agriculture," Palmer added.

BEE PROBLEMS

Without bees and other pollinators, agriculture would be in trouble. Bees are highly efficient at transferring pollen from male to female plants so they can reproduce and create much of the food we eat.

According to the U.S. Department of Agriculture, many beekeepers in this country saw a 30 to 90 percent decline in their honeybees in 2006, an event that has since been labeled as Colony Collapse Disorder (CCD). Little explanation has been found for CCD, and it appears to have happened at other intervals over the 19th and 20th centuries.

Over time, CCD has reduced the bee population by a significant amount, and only about 2.5 million bee colonies are in existence today compared to the 5 million in the 1940s.

Palmer acknowledged that CCD was a problem nationwide, but also pointed out that there are other issues making honeybee populations vulnerable. He added that more and more bees are lost during Vermont winters and fewer pounds of honey are produced during the warm months.

According to Palmer, lack of forage material, such as clover, wildflowers, and other food for pollinators has created an issue in Vermont, and also corresponds to the loss of 150,000 acres of hay over the last decade.

"There's no pastures anymore," Palmer said. "The Champlain Valley is turning into a cornfield."

He added, "There's no honey in corn."

Palmer said he is not against farmers or anything approaching that, but pointed out that a better system needs to be found.

"There's got to be some kind of balance," he said. "Right now, it's really out of balance."

In addition, Palmer pointed out that pesticides, such as Round-Up, kill a lot of bees and other critters, and that honeybees are also battling various fungal diseases and mites coming from abroad and sourcing in the US. One such disease, American Foulbrood, can only be eradicated by burning the infected hives.

Hobby beekeepers are susceptible to diseases especially, due to buying used and potentially infected equipment and lacking the proper knowledge of what different diseases look like. The Vermont Agency of Agriculture only has one part-time inspector who can respond to reports of honeybee disease, and according to Palmer, that inspector can only reach 30 percent of the state.

"It's just a struggle," Palmer said.

A SWEETER FUTURE?

Despite the numerous obstacles, Palmer is optimistic that, somehow, some way, honeybees and beekeepers will keep going onward.

"I think we're always going to have bees," he said. "[Beekeepers are] ingenious—they'll always find a way to keep going."

Hobby beekeepers in general are on the rise in Vermont, with more than two thousand in the state. Palmer sees an increase in sustainable apiaries—using homegrown queens—as a way to help ensure the future of the state's honeybees.

"That's the answer," he said. "It's very easy to do."

Palmer has other ideas as well. "I think we have to somehow try to increase the forage," he said, indicating that any new green space in Vermont can include pollinator food, like clover, wildflowers, and other plants. "I think that'd be a good plan," he said.

Palmer also indicated that the state needs a more viable bee health inspection program in order to protect good honeybees from disease. Increasing the fee a beekeeper pays to register his or her hives in Vermont to help pay for a three-month, summertime inspector was Palmer's suggestion.

As for his personal future, Palmer said he'll keep doing what he's been doing for as long as he can.

"I'll always keep bees," he said.

How Beekeeping Found Us

by Valarie Wilson, co-owner of Heavenly Honey Apiary

Scott and Valarie Wilson of Heavenly Honey Apiary.

Early in the year 2007, I heard a radio advertisement for a Beginning Beekeeping 101 class. It was being held at the Depot Home and Gardens and the cost was only $20.00. Immediately I was interested. When I arrived at work that morning I shared with Ann, my co-worker, how cool I thought it would be to raise honeybees since our property butts up against an apple orchard on both the west and north sides. Ann agreed this could be exciting and rewarding, and she quickly added that she'd buy honey from me.

About one week lapsed and I hadn't done anything about my piqued interest, when out of the blue, my husband Scott and I were driving somewhere and he blurted out, "Do you know what I want our next hobby to be?" Not even a half of a second passed before I responded "Beekeeping!" Astonished at my answer, Scott said, "How did you know? It must have been that I stressed the word "bee." "No," I said, and went on to explain what had been on my mind.

Scott was convinced that this was a blessing from God—for both of us to be on the same page! He was so excited that we would have another hobby we could both work on together. And I was thrilled because, for whatever reason, I had been spending too much time lately pondering the arrival of 2010, when we would become empty nesters. I'd become convinced that we needed to begin a few joint activities now, so that when this day came, we wouldn't be just looking at each other saying, "Who are you?"

We had already become Boy Scout leaders, and we were working our very large vegetable garden together. Participating in church worship and weekly activities is another way we spend time together. But I know that Scouts will probably end along with our child flying from the nest, and there are many hours outside of church events. And gardening, another shared hobby, is a much shorter season in Vermont than when we were living in California, so I was already anticipating our new excursions on our tandem kayak, but now we had a new hobby to get excited about—beekeeping.

The days between registering for the beekeeping class and actually attending it seemed like months instead of weeks. February 21, 2007 finally arrived, and we spent the day with seasoned beekeeper Lynn Lang, learning until our brains were about to explode. Lynn was using words like brood box, supers, and nucs. Then there was terminology for the bees themselves: drones, workers, and queens. That was easy enough, only three types, but then he mentioned nurses and foragers; how did they fit in? Oh—did we really want to move forward with this?

With a resounding YES, we wanted into this world of husbandry that has been around for centuries, yet we knew absolutely nothing about this particular realm, except for the fact that I love to consume that sticky golden-liquid that honeybees produce.

We were assured it wasn't too late to begin that very season. We could still order our bees and purchase a hive. Again there was much to be learned. We had our Memorial Day weekend all planned, our annual camping trip with my sister Vicki and her family. Sure enough, this was the same weekend that our packaged bees would arrive. It was mandatory that they be picked up on the Saturday that we were to be on vacation. Two of our church friends, Stephanie and Belinda, both post-grad students at UVM, offered to drive north to Swanton to pick up our bees and drive them to our house. This meant that instead of a three hour drive to Swanton, Scott would only need to drive one and a half hours to our home, so that he could install the bees in their hive, and return to our campground before nightfall.

As the season progressed, we read our bee books, inspected our hive, and spotted our queen. We joined the Vermont Beekeepers Association and went to their summer meeting. During this meeting I found myself being voted in as their new librarian. We learned that our association provides monthly workshops at the North Yard and South Yard. We attended all of the rest of the North Yard workshops for the season and continued to gain knowledge about our newest hobby.

September 13th rolled around, and we were invited to a Vermont honey harvesting party at Buck Hollow Apiary, which is run by Dave and Laurie Robistow. It was a hands-on event, and Scott and I were both able to uncap the honey filled frames and hand them over to be inserted into the extractor, which spins the frames until nothing but waxcomb is left. Dave and Laurie will never know what a blessing it was to get this hands-on experience before embarking on our own harvest. Again, some local friends in Monkton who weren't keeping bees this particular season helped us by lending their hand-turned extractor, stainless-steel frame holder, and uncapping tool.

The Wilson honey harvesting day arrived on September 16, 2007, and our neighbors Russ and Robin Baker came to assist us. What a wonderful plan. They had never experienced a honey harvest before; therefore, they couldn't laugh at us if we did it all wrong. Surprisingly, it went very smoothly, and I think we all had fun doing it. Our honeybees produced forty-five pounds of honey for us our first season. We couldn't have been more thrilled.

As you can see, we didn't find beekeeping; rather, it found us. We weren't looking for it, but when Lynn's workshop was announced on the radio that day, who would have thought it would lead to such a change in our lives. Now that's the bee buzz, and I'm sticking to it!

Resources

Education and Film:

The film *Vanishing of the Bees*, narrated by Ellen Page, takes a piercing, investigative look at the economic, political, and ecological implications of the worldwide disappearance of the honeybee. Directors George Langworthy and Maryam Henein present not just a story about the mysterious phenomenon known as Colony Collapse Disorder but a platform of solutions, encouraging audiences to be the change they want to see in the world. Check out the film at www.vanishingbees.com.

Another great film, *More Than Honey*, is one of my favorite films on the subject. It opens with a Swiss beekeeper hunting for honey in the wild. Over the past fifteen years, numerous colonies of bees have been decimated throughout the world, but the causes of this disaster remain unknown. Depending on the world region, 50 to 90 percent of all local bees have disappeared, and this epidemic is still spreading from beehive to beehive—all over the planet. Everywhere, the same scenario is repeated: billions of bees leave their hives, never to return. No bodies are found in the immediate surroundings, and no visible predators can be located.

In the US, the latest estimates suggest that a total of 1.5 million (out of 2.4 million total beehives) have disappeared across twenty-seven states. In Germany, according to the national beekeepers association, one fourth of all colonies have been destroyed, with related losses reaching up to 80 percent on some farms. The same phenomenon has been observed in Switzerland, France, Italy, Portugal, Greece, Austria, Poland and England, where this syndrome has been nicknamed "the Mary Celeste Phenomenon," after a ship whose crew vanished in 1872.

Scientists have found a name for the phenomenon that matches its scale—"colony collapse disorder"—and they have good reason to be worried: 80 percent of plant species require bees for pollination. Without bees to pollinate, fruits and vegetables could disappear from the face of the Earth. Apis mellifera (the honeybee) appeared on Earth sixty million years before man and is as indispensable to the economy as it is to man's survival.

Should we blame pesticides or even medication used to combat them? Maybe look at parasites such as varroa mites? New viruses? Travelling stress? The multiplication of electromagnetic waves disturbing the magnetite nanoparticles found in the bees' abdomens? So far, it looks like a combination of all these agents has been responsible for the weakening of the bees' immune defenses.

Check out the film at www.morethanhoneyfilm.com.

Solitary Bees Film
(www.helpabee.org/ca-native-bee-documentary.html)
They are crucial pollinators, yet are little known or conserved. This film showcases the fascinating behavior and value of the UK's solitary bees. We follow a variety of different species through their struggles to find resources, avoid death, and create new life.

Our main aim is for this film to not only entertain, but to be used as a free educational resource. If you would like to use *The Solitary Bees* in an educational setting, please contact Team Candiru for a free HD download. They are making a film about the Bees of California.

Online and Other Sources:

The Backyard Beekeepers Association
(www.backyardbeekeepers.com)

BeeMaid
(www.beemaid.com)
In the early 1950s, a few Western Canadian beekeepers had a dream—a vision to form an organization, owned by Canadian beekeepers, that would have the capability to sell their quality Canadian honey throughout the world. Bee Maid Honey commenced operation in 1954 when the Manitoba and Saskatchewan Honey Co-Operatives agreed to market all their honey jointly. In 1961, the

Alberta Honey Co-op participated with the Manitoba and Saskatchewan Co-Ops through Bee Maid Honey in developing the export market, and in 1962 began full participation in both the domestic and export markets.

Beesource Beekeeping
(www.beesource.com)
The Beesource Beekeeping website was started in 1997 by a hobbyist beekeeper and became an online community for beekeepers and beekeeping in 1999. It has experienced organic, word of mouth grassroots growth ever since. Today, Beesource.com has over twenty-three thousand registered members and is the most active online beekeeping community of its kind in the world.

The Bumblebee Conservation Trust
(www.bumblebeeconservation.org)
The Bumblebee Conservation Trust was established because of serious concerns about the "plight of the bumblebee." In the last eighty years our bumblebee populations have crashed. Two species have become nationally extinct and several others have declined dramatically. Bumblebees are familiar and much-loved insects that pollinate our crops and wildflowers, so people are rightly worried. We have a vision for a different future in which our communities and countryside are rich in bumblebees and colourful flowers, supporting a diversity of wildlife and habitats for everyone to enjoy. A growing number of committed supporters are helping our small team of staff make a big difference. They have over seven thousand members and are growing fast.

Judi Harvest
(www.judiharvest.com)
American glass artist Judi Harvest, whose work focuses on the decimation of bee populations, is keeping bee hives and making honey near her glass-blowing studio in Murano, Italy. She lives in New York City where she keeps bees as well.

Dr. Reese Halter, *The Incomparable Honeybee*
(www.drreese.com/infoBooks)
From Dr. Reese Halter comes a remarkable, concise account of the honeybees that have profoundly shaped our planet for the past 110

million years. They are the most important group of flower-visiting animals, pollinating more multi-billion-dollar crops and plants than any other living group. Since prehistoric times humans and honeybees have been inextricably linked. This book is rich with interesting and humbling facts: bees can count, they can vote, and honey has potent medicinal properties, able to work as an anti-inflammatory, antibacterial, antifungal, antioxidant, even an antiseptic. The fate of the bees, whose numbers have been beleaguered most recently by colony collapse disorder, lies firmly in the hands of humankind. As such, it is our job to ensure their health, protect the habitats within which they live, and communicate to others the vital link that human society shares with the remarkable honeybee.

Heavenly Honey Apiary
(www.vtbeekeeper.com)
This apiary is owned and operated by Scott and Valarie Wilson. In the beekeeping world we are known as sideliners. What started as a hobby has now grown into a small Vermont honey business.

The National Honey Board
(www.honey.com)
The National Honey Board (NHB) is an industry-funded agriculture promotion group that works to educate consumers about the benefits and uses for honey and honey products through research, marketing, and promotional programs. The Board's work, funded by an assessment of one cent per pound on domestic and imported honey, is designed to increase the awareness and usage of honey by consumers, the foodservice industry, and food manufacturers. The ten-member-Board, appointed by the U.S. Secretary of Agriculture, represents producers (beekeepers), packers, importers and a marketing cooperative.

The Vermont Beekeepers Association
(www.vermontbeekeepers.org)
Since 1886 the VBA has promoted the general welfare of Vermont's Honey Industry, while sustaining a friendly body of unity among the state's beekeepers.

The Vermont Beekeepers Association, a 501(c)(3) nonprofit organization, represents hundreds of beekeepers that raise bees for the love and honey. We're as diverse as the 246 towns in Vermont, but are unified in our fascination with and affection for bees. Most of us are hobbyists, but there are some "side liners" who try to make a bit of extra income from their 25–200 hives as well as a handful of full-time professionals

The Xerces Society for Invertebrate Conservation (www.xerces.org)

The Xerces Society is a nonprofit organization that protects wildlife through the conservation of invertebrates and their habitat. For over forty years, the Society has been at the forefront of invertebrate protection worldwide, harnessing the knowledge of scientists and the enthusiasm of citizens to implement conservation programs.

Glossary

Beeswax: Waxy material produced by worker bees and used to build combs.

Drones: Male bees, whose main function in the colony is to fertilize the queen. Drones make up a very small percentage of the total colony. In the Autumn drones are expelled from the hive by the female worker bees.

Foundation: Thin sheets of beeswax imprinted with a pattern of honey comb. The beekeeper installs these sheets into wooden frames as "starters" for the bees in making uniform combs.

Frames: The removable wooden structures which are placed in the hive. The bees build their comb within these frames. The removable quality allows the beekeeper to easily inspect the colony.

Hive Bodies: The first one or two wooden boxes of the colony. The hive bodies contain the brood nest of the colony.

Larva: The grub-like, immature form of the bee, after it has developed from the egg and before it has gone into the pupa stage.

Nectar: Sweet fluid produced by flowers is 60 percent water and 40 percent solids. This is collected by the bees and converted into honey at 17–18 percent moisture content.

Pollen: Very small dust-like grain produced by flowers. These are the male germ cells of the plant.

Propolis: Sticky, brownish gum gathered by bees from trees and buds and used to seal cracks and drafts in the hive. Also called "bee-glue."

Pupa: The immature form of the bee (following the larval stage) while changing into the adult form.

Queen: A completely developed female bee (with functioning ovaries) who lays eggs and serves as the central focus of the colony. There is only one queen in a colony of bees. A queen's productive life span is 2–3 years.

Royal Jelly: The milky white secretion of young nurse bees. It is used to feed the queen throughout her life, and is given to worker and drone larvae only during their early larval lives.

Super: The supplementary wooden boxes placed on top of the hive body to expand the size of the colony, and to provide for storage of surplus honey.

Supercedure: When a colony with an old or failing queen rears a daughter to replace her.

Workers: Completely developed female bees that do have developed ovaries and do not not normally lay eggs. They gather pollen and nectar and convert the nectar to honey. A worker's life expectancy is only several weeks during the active summer months. However, they can live for many months during the relatively inactive winter period.

Bibliography

"A Simple Harvest." *Back Yard Hive*. N.d. http://www.backyardhive.com/ articles_on_ beekeeping/articles_on_beekeeping/a_simple_harvest/.

"Ancient Egyptian Bee-keeping." N.d. http://www.reshafim.org.il/ad/ egypt/timelines/topics/beekeeping.htm.

"Beekeeping Equipment." *Bee Thinking*. N.d. http://www.beethinking.com/ collections/tools-and-equipment.

Blackiston, Howland. "The Langstroth Hive: How to Build Your Own." *Building Beehives for Dummies*. N.d. http://www.dummies.com/how-to/ content/cut-list-for-the-tenframe-langstroth-hive.html.

Boyle, Alan. "Bee-Killing Pesticide Found in Garden Store Plants." CNBC News. Last modified August 15, 2013. http://www.cnbc.com/ id/100965440.

Brown, Heather. "Beekeeping 101: Where to Get Bees." The Old Farmer's Almanac. Last modified May 2, 2012. http://www.almanac.com/blog/ beekeeping/beekeeping-101-where-get-bees.

Burlew, Rusty. "Beekeeping Essentials for the Beginner: Langstroth Hives." Honey Bee Suite. Last modified January 6, 2011. Web. http://www. honeybeesuite.com/beekeeping-essentials-for-the-beginner/.

Burlew, Rusty. "The Types of Hive Tools." Honey Bee Suite. Last modified April 12, 2010. http://www.honeybeesuite.com/the-types-of-hive-tools/.

Caldeira, John. "Backyard Beekeeping." John's Beekeeping Notebook. N.d. http://outdoorplace. org/beekeeping/citybees.htm.

"How to Build Your Own DIY Top Bar Beehive." *Remove and Replace*. Last modified April 9, 2014. http://removeandreplace.com/2014/04/09/ how-to-build-your-own-diy-top-bar-beehive/.

Mandal, Manisha D. and Shyamapada M. "Honey: Its Medicinal Property and Antibacterial Activity." *Asian Pacific Journal of Tropical Biomedicine* 1, no. 2 (2011): 154–160. http://www.ncbi.nlm.nih.gov/pmc/articles/ PMC3609166/.

Marks, Katie. "Where to Put Your Beehive: More Bees, Please!" *Networx*. Last modified January 16, 2014. http://www.networx.com/article/where-to-put-your-beehive.

Miller, Julia. "Location, Location, Location: Siting Beehives." *Mother Earth News*. Last modified January 15, 2014. http://www.motherearthnews.com/homesteading-and-livestock/where-to-put-your-hives-zbcz1401.aspx.

Pica, Erich. "Gardeners Beware: 'Bee-friendly' Plants May Be Poisoning Your Garden." *Friends of the Earth*. Last modified August 14, 2013. http://www.foe.org/news/archives/gardeners-beware.

Smith, Mark. "Apiculture and Permaculture: Keeping Bees." Permaculture. N.d. http://www.permaculture.co.uk/articles/apiculture-and-permaculture-keeping-bees.

Staton, Stephanie. "How to Harvest Honey." Urban Farm Online. N.d. http://www.urbanfarmonline.com/urban-livestock/bee-keeping/harvest-honey.aspx.

Tromp, David. "Traditional Top Bar Hives." Beesource Beekeeping. N.d. http://www.beesource.com/resources/elements-of-beekeeping/alternative-hive-designs/traditional-top-bar-hives-david-tromp/.

Williamson, Lindsay. "Langstroth, Top-Bar or Warre?: Choose the Right Hive for You and Your Bees." Mother Earth News. Last modified October 25, 2013. http://www.motherearthnews.com/homesteading-and-livestock/langstroth-top-bar-or-warre-zbcz1310.aspx.

INDEX

AFTERWORD

"Nobody disputes the role of dogs as man's best friend, but a convincing argument can also be made for the honey bee."
—Martin Elkort, *The Secret Life of Food*

BEEKEEPING: NEEDED NOW MORE THAN EVER

According to an article on Alternet.com by environment/food editor Reynard Loki, a quarter of everything we eat—from apples to zucchini—

depends on pollination, which means it depends on bees pollinating the crops. But beekeepers from around the world are reporting the worst pandemic of a bee population collapse ever recorded. Chemical-giant Bayer, which markets aspirin and Alka Seltzer, has toxic agricultural pesticides linked to this unprecedented bee die-off. The EPA has the power to ban the pesticides, but so far has bowed to pressure from Bayer. Our nation's food supply is at risk without our bees!

It is time to stand up for our friends, the bees, extraordinary pollinators who are an integral part of our survival. I plan to learn from my neighbor, Gordon Fisher, and come spring of next year, I will have one fresh, new Langstroth hive in my backyard. I have a feeling the apples in the orchard down the road are going to appreciate this new item as my property abuts an organic orchard! I am going to plant flowers around my property that will attract bees, too, and work tirelessly to stop GMO-modified plants and more pesticides from being applied to corn, soybeans, and other big agri-crops.

"You are what you eat, and no one wants to be eating any poison. Plant native yellow and blue flowers in large groups, and don't use ANY herbicides in your garden."

—Dr. Reese Halter, biologist

Acknowledgments

Writing this book was an amazing experience. I couldn't have done it without help from a number of people and I would like to thank them, as follows: my high school intern, Lindsey Vachon, was involved almost every step of the way. She and I worked on the book together and without her help I don't think it would be such a great resource. I'm very grateful for the work Lindsey did to help this book take shape. I also had help from all of the beekeepers I interviewed. They are an amazing group of people, without whom this book would be more of a how-to book instead of the fun, informative guide filled with "stories from the hives." Thank you ALL for your contributions! Scott and Valerie Wilson, who run Vermont's Heavenly Honey Apiary, are an inspiration to me as a newbie beekeeper. Scott wrote the beautiful Foreword and Valerie supported this book from the beginning! Other help I am grateful to have received came from friends/colleagues, Sierra Dickey, Rose Alexandre Leach, Danielle Gyger, Nancy Miller and many others who have supported my foray into sustainable living! I am especially grateful to my editor, Abigail Gehring, who has been my number one supporter since I proposed my first book to her—she is also a fellow Vermonter who shares my passion for natural living. The team at Skyhorse and Good Books is, as always, a great group and I am grateful for all their help! Happy beekeeping!

NOTES

NOTES

NOTES

NOTES

NOTES

NOTES